LONDON MATHEMATICAL SOCIETY LECTURE NOTE SERIES

Managing Editor: Professor N.J. Hitchin, Mathematical Institute,
University of Oxford, 24–29 St Giles, Oxford OX1 3LB, United Kingdom

The titles below are available from booksellers, or, in case of difficulty, from Cambridge University Press.

46   *p*-adic Analysis: a short course on recent work,   N. KOBLITZ
59   Applicable differential geometry,   M. CRAMPIN & F.A.E. PIRANI
66   Several complex variables and complex manifolds II,   M.J. FIELD
76   Topological topics,   I.M. JAMES (ed)
88   FPF ring theory,   C. FAITH & S. PAGE
90   Polytopes and symmetry,   S.A. ROBERTSON
96   Diophantine equations over function fields,   R.C. MASON
97   Varieties of constructive mathematics,   D.S. BRIDGES & F. RICHMAN
99   Methods of differential geometry in algebraic topology,   M. KAROUBI & C. LERUSTE
100 Stopping time techniques for analysts and probabilists,   L. EGGHE
104 Elliptic structures on 3-manifolds,   C.B. THOMAS
105 A local spectral theory for closed operators,   I. ERDELYI & WANG SHENGWANG
107 Compactification of Siegel moduli schemes,   C.-L. CHAI
109 Diophantine analysis,   J. LOXTON & A. VAN DER POORTEN (eds)
113 Lectures on the asymptotic theory of ideals,   D. REES
116 Representations of algebras,   P.J. WEBB (ed)
119 Triangulated categories in the representation theory of finite-dimensional algebras,   D. HAPPEL
121 Proceedings of *Groups - St Andrews 1985*,   E. ROBERTSON & C. CAMPBELL (eds)
128 Descriptive set theory and the structure of sets of uniqueness,   A.S. KECHRIS & A. LOUVEAU
130 Model theory and modules,   M. PREST
131 Algebraic, extremal & metric combinatorics,   M.-M. DEZA, P. FRANKL & I.G. ROSENBERG (eds)
138 Analysis at Urbana, II,   E. BERKSON, T. PECK, & J. UHL (eds)
139 Advances in homotopy theory,   S. SALAMON, B. STEER & W. SUTHERLAND (eds)
140 Geometric aspects of Banach spaces,   E.M. PEINADOR & A. RODES (eds)
141 Surveys in combinatorics 1989,   J. SIEMONS (ed)
144 Introduction to uniform spaces,   I.M. JAMES
145 Cohen-Macaulay modules over Cohen-Macaulay rings,   Y. YOSHINO
148 Helices and vector bundles,   A.N. RUDAKOV *et al*
149 Solitons, nonlinear evolution equations and inverse scattering,   M. ABLOWITZ & P. CLARKSON
150 Geometry of low-dimensional manifolds 1,   S. DONALDSON & C.B. THOMAS (eds)
151 Geometry of low-dimensional manifolds 2,   S. DONALDSON & C.B. THOMAS (eds)
152 Oligomorphic permutation groups,   P. CAMERON
153 L-functions and arithmetic,   J. COATES & M.J. TAYLOR (eds)
155 Classification theories of polarized varieties,   TAKAO FUJITA
158 Geometry of Banach spaces,   P.F.X. MÜLLER & W. SCHACHERMAYER (eds)
159 Groups St Andrews 1989 volume 1,   C.M. CAMPBELL & E.F. ROBERTSON (eds)
160 Groups St Andrews 1989 volume 2,   C.M. CAMPBELL & E.F. ROBERTSON (eds)
161 Lectures on block theory,   BURKHARD KÜLSHAMMER
163 Topics in varieties of group representations,   S.M. VOVSI
164 Quasi-symmetric designs,   M.S. SHRIKANDE & S.S. SANE
166 Surveys in combinatorics, 1991,   A.D. KEEDWELL (ed)
168 Representations of algebras,   H. TACHIKAWA & S. BRENNER (eds)
169 Boolean function complexity,   M.S. PATERSON (ed)
170 Manifolds with singularities and the Adams-Novikov spectral sequence,   B. BOTVINNIK
171 Squares,   A.R. RAJWADE
172 Algebraic varieties,   GEORGE R. KEMPF
173 Discrete groups and geometry,   W.J. HARVEY & C. MACLACHLAN (eds)
174 Lectures on mechanics,   J.E. MARSDEN
175 Adams memorial symposium on algebraic topology 1,   N. RAY & G. WALKER (eds)
176 Adams memorial symposium on algebraic topology 2,   N. RAY & G. WALKER (eds)
177 Applications of categories in computer science,   M. FOURMAN, P. JOHNSTONE & A. PITTS (eds)
178 Lower K- and L-theory,   A. RANICKI
179 Complex projective geometry,   G. ELLINGSRUD *et al*
180 Lectures on ergodic theory and Pesin theory on compact manifolds,   M. POLLICOTT
181 Geometric group theory I,   G.A. NIBLO & M.A. ROLLER (eds)
182 Geometric group theory II,   G.A. NIBLO & M.A. ROLLER (eds)
183 Shintani zeta functions,   A. YUKIE
184 Arithmetical functions,   W. SCHWARZ & J. SPILKER
185 Representations of solvable groups,   O. MANZ & T.R. WOLF
186 Complexity: knots, colourings and counting,   D.J.A. WELSH
187 Surveys in combinatorics, 1993,   K. WALKER (ed)
188 Local analysis for the odd order theorem,   H. BENDER & G. GLAUBERMAN
189 Locally presentable and accessible categories,   J. ADAMEK & J. ROSICKY
190 Polynomial invariants of finite groups,   D.J. BENSON
191 Finite geometry and combinatorics,   F. DE CLERCK *et al*
192 Symplectic geometry,   D. SALAMON (ed)
193 Independent random variables and rearrangement invariant spaces,   M. BRAVERMAN
194 Arithmetic of blowup algebras,   WOLMER VASCONCELOS
195 Microlocal analysis for differential operators,   A. GRIGIS & J. SJÖSTRAND
196 Two-dimensional homotopy and combinatorial group theory,   C. HOG-ANGELONI *et al*

198   The algebraic characterization of geometric 4-manifolds, J.A. HILLMAN
199   Invariant potential theory in the unit ball of C^n, MANFRED STOLL
200   The Grothendieck theory of dessins d'enfant, L. SCHNEPS (ed)
201   Singularities, JEAN-PAUL BRASSELET (ed)
202   The technique of pseudodifferential operators, H.O. CORDES
203   Hochschild cohomology of von Neumann algebras, A. SINCLAIR & R. SMITH
204   Combinatorial and geometric group theory, A.J. DUNCAN, N.D. GILBERT & J. HOWIE (eds)
205   Ergodic theory and its connections with harmonic analysis, K. PETERSEN & I. SALAMA (eds)
207   Groups of Lie type and their geometries, W.M. KANTOR & L. DI MARTINO (eds)
208   Vector bundles in algebraic geometry, N.J. HITCHIN, P. NEWSTEAD & W.M. OXBURY (eds)
209   Arithmetic of diagonal hypersurfaces over finite fields, F.Q. GOUVÊA & N. YUI
210   Hilbert C*-modules, E.C. LANCE
211   Groups 93 Galway / St Andrews I, C.M. CAMPBELL et al (eds)
212   Groups 93 Galway / St Andrews II, C.M. CAMPBELL et al (eds)
214   Generalised Euler-Jacobi inversion formula and asymptotics beyond all orders, V. KOWALENKO et al
215   Number theory 1992–93, S. DAVID (ed)
216   Stochastic partial differential equations, A. ETHERIDGE (ed)
217   Quadratic forms with applications to algebraic geometry and topology, A. PFISTER
218   Surveys in combinatorics, 1995, PETER ROWLINSON (ed)
220   Algebraic set theory, A. JOYAL & I. MOERDIJK
221   Harmonic approximation, S.J. GARDINER
222   Advances in linear logic, J.-Y. GIRARD, Y. LAFONT & L. REGNIER (eds)
223   Analytic semigroups and semilinear initial boundary value problems, KAZUAKI TAIRA
224   Computability, enumerability, unsolvability, S.B. COOPER, T.A. SLAMAN & S.S. WAINER (eds)
225   A mathematical introduction to string theory, S. ALBEVERIO, J. JOST, S. PAYCHA, S. SCARLATTI
226   Novikov conjectures, index theorems and rigidity I, S. FERRY, A. RANICKI & J. ROSENBERG (eds)
227   Novikov conjectures, index theorems and rigidity II, S. FERRY, A. RANICKI & J. ROSENBERG (eds)
228   Ergodic theory of Z^d actions, M. POLLICOTT & K. SCHMIDT (eds)
229   Ergodicity for infinite dimensional systems, G. DA PRATO & J. ZABCZYK
230   Prolegomena to a middlebrow arithmetic of curves of genus 2, J.W.S. CASSELS & E.V. FLYNN
231   Semigroup theory and its applications, K.H. HOFMANN & M.W. MISLOVE (eds)
232   The descriptive set theory of Polish group actions, H. BECKER & A.S. KECHRIS
233   Finite fields and applications, S. COHEN & H. NIEDERREITER (eds)
234   Introduction to subfactors, V. JONES & V.S. SUNDER
235   Number theory 1993–94, S. DAVID (ed)
236   The James forest, H. FETTER & B. GAMBOA DE BUEN
237   Sieve methods, exponential sums, and their applications in number theory, G.R.H. GREAVES et al
238   Representation theory and algebraic geometry, A. MARTSINKOVSKY & G. TODOROV (eds)
239   Clifford algebras and spinors, P. LOUNESTO
240   Stable groups, FRANK O. WAGNER
241   Surveys in combinatorics, 1997, R.A. BAILEY (ed)
242   Geometric Galois actions I, L. SCHNEPS & P. LOCHAK (eds)
243   Geometric Galois actions II, L. SCHNEPS & P. LOCHAK (eds)
244   Model theory of groups and automorphism groups, D. EVANS (ed)
245   Geometry, combinatorial designs and related structures, J.W.P. HIRSCHFELD et al
246   p-Automorphisms of finite p-groups, E.I. KHUKHRO
247   Analytic number theory, Y. MOTOHASHI (ed)
248   Tame topology and o-minimal structures, LOU VAN DEN DRIES
249   The atlas of finite groups: ten years on, ROBERT CURTIS & ROBERT WILSON (eds)
250   Characters and blocks of finite groups, G. NAVARRO
251   Gröbner bases and applications, B. BUCHBERGER & F. WINKLER (eds)
252   Geometry and cohomology in group theory, P. KROPHOLLER, G. NIBLO, R. STÖHR (eds)
253   The q-Schur algebra, S. DONKIN
254   Galois representations in arithmetic algebraic geometry, A.J. SCHOLL & R.L. TAYLOR (eds)
255   Symmetries and integrability of difference equations, P.A. CLARKSON & F.W. NIJHOFF (eds)
256   Aspects of Galois theory, HELMUT VÖLKLEIN et al
257   An introduction to noncommutative differential geometry and its physical applications 2ed, J. MADORE
258   Sets and proofs, S.B. COOPER & J. TRUSS (eds)
259   Models and computability, S.B. COOPER & J. TRUSS (eds)
260   Groups St Andrews 1997 in Bath, I, C.M. CAMPBELL et al
261   Groups St Andrews 1997 in Bath, II, C.M. CAMPBELL et al
263   Singularity theory, BILL BRUCE & DAVID MOND (eds)
264   New trends in algebraic geometry, K. HULEK, F. CATANESE, C. PETERS & M. REID (eds)
265   Elliptic curves in cryptography, I. BLAKE, G. SEROUSSI & N. SMART
267   Surveys in combinatorics, 1999, J.D. LAMB & D.A. PREECE (eds)
268   Spectral asymptotics in the semi-classical limit, M. DIMASSI & J. SJÖSTRAND
269   Ergodic theory and topological dynamics, M.B. BEKKA & M. MAYER
270   Analysis on Lie Groups, N.T. VAROPOULOS & S. MUSTAPHA
271   Singular perturbations of differential operators, S. ALBEVERIO & P. KURASOV
272   Character theory for the odd order function, T. PETERFALVI
273   Spectral theory and geometry, E.B. DAVIES & Y. SAFAROV (eds)
274   The Mandlebrot set, theme and variations, TAN LEI (ed)
276   Singularities of plane curves, E. CASAS-ALVERO
279   Topics in symbolic dynamics and applications, F. BLANCHARD, A. MAASS & A. NOGUEIRA (eds)
281   Explicit birational geometry of 3-folds, ALESSIO CORTI & MILES REID (eds)

London Mathematical Society Lecture Note Series. 269

Ergodic Theory and Topological Dynamics of Group Actions on Homogeneous Spaces

M. Bachir Bekka
Université de Metz

Matthias Mayer
KPMG, Munich

CAMBRIDGE
UNIVERSITY PRESS

CAMBRIDGE UNIVERSITY PRESS
Cambridge, New York, Melbourne, Madrid, Cape Town,
Singapore, São Paulo, Delhi, Mexico City

Cambridge University Press
The Edinburgh Building, Cambridge CB2 8RU, UK

Published in the United States of America by Cambridge University Press, New York

www.cambridge.org
Information on this title: www.cambridge.org/9780521660303

First published 2000

A catalogue record for this publication is available from the British Library

ISBN 978-0-521-66030-3 Paperback

Contents

Preface   vii

I. Ergodic Systems   1

§1 Examples and Basic Results   1
§2 Ergodic Theory and Unitary Representations   13
§3 Invariant Measures and Unique Ergodicity   30

II. The Geodesic Flow of Riemannian Locally Symmetric
Spaces   36

§1 Some Hyperbolic Geometry   38
§2 Lattices and Fundamental Domains   42
§3 The Geodesic Flow of Compact Riemann Surfaces   57
§4 The Geodesic Flow on Riemannian Locally Symmetric Spaces   62

III. The Vanishing Theorem of Howe and Moore   80

§1 Howe–Moore's Theorem   81
§2 Moore's Ergodicity Theorems   89
§3 Counting Lattice Points in the Hyperbolic Plane   93
§4 Mixing of All Orders   98

IV. The Horocycle Flow   110

§1 The Horocycle Flow of a Riemann Surface   111
§2 Proof of Hedlund's Theorem – Cocompact Case   116
§3 Classification of Invariant Measures   120
§4 Equidistribution of Horocycle Orbits   128

V. Siegel Sets, Mahler's Criterion and Margulis' Lemma   139

§1 Siegel Sets in $SL(n, \mathbb{R})$ 139
§2 $SL(n, \mathbb{Z})$ is a lattice in $SL(n, \mathbb{R})$ 144
§3 Mahler's Criterion   146
§4 Reduction of Positive Definite Quadratic Forms   148
§5 Margulis' Lemma   150

VI. An Application to Number Theory:
Oppenheim's Conjecture   161

§1 Oppenheim's Conjecture   162
§2 Proof of the Theorem – Preliminaries   163
§3 Existence of Minimal Closed Subsets   172
§4 Orbits of One-Parameter Groups of Unipotent
Linear Transformations   177
§5 Proof of the Theorem – Conclusion   179
§6 Ratner's Results on the Conjectures of Raghunathan,
Dani and Margulis   184

Bibliography   189

Index   198

Preface

These notes are based on lectures given in 1994 by the first author at a Summer School in Tuczno (Poland) and at the University of Metz.

The purpose is to give a quick introduction to ergodic theory, to study interesting examples as illustration and to present some recent and spectacular developments in topological dynamics of group actions. More precisely, the focus will be on the following two types of systems of a geometrical–algebraic nature:

The geodesic flow on the unit tangent bundle of a locally symmetric space and unipotent actions on homogeneous spaces. Classical examples are the geodesic flow and the horocyclic flow on the unit tangent bundle of a compact Riemann surface of constant negative curvature. These flows are among the most studied dynamical systems. Of particular interest are their ergodic (or mixing) properties and the asymptotic behaviour of their orbits.

Unipotent actions on homogeneous spaces enjoy remarkable regularity properties. A striking illustration of this regularity is Hedlund's minimality theorem: for any lattice Γ in $G = \mathrm{SL}(2, \mathbb{R})$, any orbit in the homogeneous space $\Gamma \setminus G$ under a unipotent subgroup of $\mathrm{SL}(2, \mathbb{R})$ is either periodic or dense. Such actions have close connections to problems in Number Theory. For instance, one of the most spectacular application of unipotent actions is the solution in 1987 by Margulis of Oppenheim's conjecture which was open for more than 40 years (see [Ma3]). Margulis' proof is based on the study of the dynamical properties of the action of unipotent one-parameter subgroups of $\mathrm{SO}(2,1)$ on $\mathrm{SL}(3, \mathbb{R}) / \mathrm{SL}(3, \mathbb{Z})$.

In this context, Raghunathan as well as Dani and Margulis formulated in the eighties general conjectures about closure and distribution of orbits of unipotent flows. These so-called Raghunathan conjectures were settled at the beginning of the nineties by Ratner ([Ra2]–[Ra6]).

In theses notes, the emphasis will be put on the group theoretical point of view, and, whenever possible, on the rôle of unitary group representations. So, ergodicity and mixing will be seen to be special instances of the powerful Howe–Moore's theorem about the asymptotic behaviour of matrix coefficients of unitary representations of semisimple groups.

Group representations enter the picture in the following way. A compact Riemann surface Σ of genus $g \geq 2$ has the real hyperbolic plane $\mathbf{H}^2$ as the universal covering space and may therefore be identified with the orbit space $\Gamma \setminus \mathbf{H}^2$ for a discrete cocompact subgroup of $\mathrm{PSL}(2, \mathbb{R})$. In this way, the geodesic flow and the horocycle flow on the unit tangent bundle $T^1(\Sigma)$ of Σ may be identified with flows on the homogeneous space $\Gamma \setminus \mathrm{PSL}(2, \mathbb{R})$ given by right translations by elements from the subgroup A and N of the diagonal and of the unipotent upper triangular matrices in $\mathrm{PSL}(2, \mathbb{R})$, respectively. Now, the constant functions on $T^1(\Sigma)$ are certainly invariant under A (or N). Ergodicity means that there are no other

A-invariant (or N-invariant) functions in $L^2(T^1(\Sigma))$, a space carrying a natural unitary representation of PSL(2, $\mathbb{R}$).

This point of view has first been adopted by Gelfand and Fomin [GF] in the fifties. Their methods were based on the explicit description of the unitary dual of SL(2, $\mathbb{R}$). In fact, such a description is not necessary for that purpose (and not always available for other groups). Mautner [Mau] treated the geodesic flow for all locally symmetric spaces in this spirit. Subsequently, Moore [Mo1] proved his celebrated ergodicity theorem, a very general result valid for all semisimple Lie groups. This, in turn, is a special case of the Howe–Moore theorem about the vanishing at infinity of matrix coefficients of unitary representations.

Here are the essential features of these notes:

(i) We give a complete proof of Howe–Moore's vanishing theorem alluded to above. Using root theory, we first reduce to the case of SL(2, $\mathbb{R}$). For this group, we follow the elementary and elegant proof from [HT]. Moore's ergodicity theorems are deduced from the vanishing theorem.

(ii) We give Mautner's characterization of ergodicity of the geodesic flow for locally symmetric spaces.

(iii) Following Ratner [Ra7], a proof of Raghanuthan's conjectures for the case SL(2, $\mathbb{R}$) is given. More precisely, the invariant measures under the horocyclic flow are classified. This classification was first obtained by Furstenberg [Fu1] in the case of cocompact lattices and by Dani [Da1] for general lattices. Moreover, equidistribution of the orbits of the horocyclic flow is established. This is Dani–Smillie's equidistribution theorem [DS] which may be viewed as a quantitative refinement of the minimality theorem of Hedlund mentioned above. Ratner's results which are valid in much greater generality are beyond the scope of these notes. However, the case of SL(2, $\mathbb{R}$) presented here already contains many crucial ideas involved in the general situation.

(iv) We give an elementary, complete proof of Oppenheim's conjecture. This conjecture says that, if Q is a nondegenerate indefinite form on $\mathbb{R}^n$, $n \geq 3$, and if Q is not a multiple of a rational form, then $Q(\mathbb{Z}^n)$ is dense in $\mathbb{R}$. Our treatment is based on the article [DM3] by Dani and Margulis. We tried to emphasize the common features of this proof with Hedlund's minimality theorem

Some other features are worth mentioning. In view of the proof of Oppenheim's conjecture, we discuss the construction of Siegel sets in $SL(n, \mathbb{R})$, proving Mahler's compactness criterion, and give the proof of Margulis' lemma about recurrence of orbits of unipotent one-parameter groups in SL(n, $\mathbb{R}$)/SL(n, $\mathbb{Z}$). As an application of mixing, we establish – following [EM] – an asymptotic formula for the number of lattice points in a ball in the hyperbolic plane. We also discuss Mozes' result about mixing of all orders of actions of semisimple Lie groups, as well as Ledrappier's counterexample of a mixing action of $\mathbb{Z}^2$ which is not mixing of all orders.

Description of the contents

Here is an overview about the contents of these notes:

Chapter I is devoted to some general facts about ergodic systems: definitions, first examples, classical theorems such as Poincaré's recurrence theorem, the ergodic theorems of von Neumann and of Birkhoff. We borrowed from M.S. Keane ([BKS]) a particularly elegant proof for Birkhoff's ergodic theorem. Special attention is paid to stating the different notions of mixing in representation theoretic terms. Finally, we present some facts about the existence of invariant measures for group actions and study the question of when ergodic measures are supported on orbits.

Chapter II deals with our first main example, the geodesic flow on a locally symmetric space. We treat in great detail the standard facts about Fuchsian groups and their fundamental domains (Dirichlet regions). We describe the geodesic flow on a Riemann surface of negative constant curvature and, more generally, on a locally symmetric Riemannian space in group theoretic terms. Mautner's rank one criterion for ergodicity is discussed.

In Chapter III, we give the complete proof of Howe–Moore's vanishing theorem. We discuss some examples in connection with Moore's duality theorem. As an application, we give the asymptotic formula for the lattice point problem in the hyperbolic plane. We discuss Ledrappier's counterexample as well as Mozes' result on mixing of all orders of Lie group actions.

Chapter IV deals with the horocycle flow. We first discuss Hedlund's minimality theorem in the case of a compact surface. Following [Ra7], we then prove the classification result of invariant measures as well as Dani–Smillie's equidistribution theorem.

In Chapter V, we discuss in length the construction of Siegel sets in $SL(n, \mathbb{R})$, showing that $SL(n, \mathbb{Z})$ is a lattice and proving Mahler's compactness criterion. This chapter contains also the complete proof of Margulis' lemma.

In Chapter VI, a strengthening – due to Dani and Margulis – of Oppenheim's conjecture is proved according to which the set of values taken by an indefinite irrational form in $n \geq 3$ variables on the primitive integer vectors in $\mathbb{Z}^n$ is dense.

Prerequisites

Concerning the prerequisites, we only assume the reader to be familiar with the elementary facts from functional analysis, measure theory and Lie theory. For convenience, more elaborate notions and results – such as the Cartan decomposition of semisimple Lie groups – are shortly recalled when they are needed. In the concrete examples we have in mind, such results correspond usually to elementary facts which are easy to prove. These notes are intended to be accessible for mathematicians and graduate students with various backgrounds.

Acknowledgment

The first author is grateful to the participants and organizers of the Tuczno Summer School and to the students and colleagues who attended his course in Metz. He is also grateful to Martine Babillot, Pierre de la Harpe, who read carefully parts of the manuscript and suggested several improvements, and to Marc Burger, S. G. Dani, and Roe Goodman for helpful discussions and remarks. Special thanks are due to the referee for his numerous comments and suggestions. We would like to acknowledge the constant inspiration we found in Etienne Ghys' nicely written Bourbaki report [Gh]. Finally, we wish to thank Roger Astley of Cambridge University Press for his help.

Chapter I

Ergodic Systems

Ergodic theory may be viewed as the study of measure (or, more generally, measure class) preserving actions of groups (or semigroups) on measure spaces.

The main examples to be treated throughout these notes arise as follows. Let G be a locally compact group, and let H, L be closed subgroups of G. The homogeneous space G/H carries a unique G-invariant measure class. Now, L acts on G/H by left translations. An interesting and important problem is to study, for specific G, H, L this action of L on G/H from a measure theoretic point of view. Usually, H is a lattice in G (see Chap. II, §2) so that G/H carries a G-invariant probability measure. So, we shall almost always deal with measure preserving actions on a probability space.

This chapter is a quick introduction to ergodic theory. We discuss mainly material which is relevant for later chapters.

Our exposition is incomplete as several important topics, such as entropy, have been omitted. Section 1 contains some standard examples of ergodic actions. In Section 2, ergodicity is formulated in terms of unitary group representations (the so-called Koopmanism). The classical ergodic theorem of von Neumann is proved and M. Keane's elegant proof of Birkhoff's ergodic theorem is reproduced. Moreover, strong mixing and weak mixing are introduced and discussed from the point of view of unitary representations. In Section 3, we state the theorem about the decomposition of general measure preserving group actions into ergodic pieces. We discuss also the existence of invariant measures for a continuous transformation of a compact space, as well as the problem of uniqueness of ergodic measures.

§1 Examples and Basic Results

Let G be a locally compact second countable group with identity e. Let (X, μ) be a measurable space with a σ-finite measure μ. Recall that a measure μ on a measure space X is *finite* if $\mu(X) < \infty$ and that μ is *σ-finite* if X is a countable union of measurable subsets of finite measure.

In what follows, all measures are assumed to be σ-finite.

1.1 Definition. An *action* of G on X is a measurable mapping

$$G \times X \to X , \ (g,x) \mapsto g\,x$$

with the following two properties:

(i) $g_1\,(g_2\,x) = (g_1\,g_2)\,x$, $e\,x = x$, for all g_1, $g_2 \in G, x \in X$, and

(ii) G preserves the measure class of μ, that is, for all measurable subsets A of X and all $g \in G$, one has $\mu(gA) = 0$ if and only if $\mu(A) = 0$. One says that μ is *quasi-invariant*.

The action of G is *ergodic* if there are no non-trivial invariant subsets of X, that is, if the following holds: if A is a measurable and G-invariant subset of X, then $\mu(A) = 0$ or $\mu(X \setminus A) = 0$.

Usually, the measure μ on X will even be *invariant* under G, that is, $\mu(gA) = \mu(A)$ for all measurable subsets A of X and all $g \in G$. However, there are interesting examples where this is not the case such as the natural action of $G = \mathrm{SL}(2, \mathbb{Z})$, the group of integer 2×2 matrices with determinant 1, on the real projective line $\mathbb{RP}^1$ (see Chap. III, Examples 2.9).

Recall that every locally compact group G has a non-zero locally finite Borel measure which is invariant under left translations and that such a measure, called *Haar measure*, is unique, up to a constant factor (see, e.g., [Wei] or [HeR]). A Borel measure μ on a locally compact space X is *locally finite* or *regular* if $\mu(K) < \infty$ for every compact subset K of X.

Very often, it is more convenient to work with functions instead of subsets. Let G be a group acting on a measure space (X, μ). A measurable function $f : X \to \mathbb{R}$ is *essentially G-invariant* if, for any $g \in G$, one has $f(gx) = f(x)$ for μ-almost all $x \in X$. The function $f : X \to \mathbb{R}$ is *G-invariant* if $f(gx) = f(x)$ for all $g \in G$ and all $x \in X$.

The following useful lemma shows that an essentially G-invariant function on X agrees almost everywhere with a G-invariant function (see [Zi], 2.2.16 Lemma). (Observe that this is obvious when G is countable.)

1.2 Lemma. *Let G be a locally compact second countable group acting on a σ-finite measure space (X, μ). Let $f : X \to \mathbb{R}$ be a measurable essentially G-invariant function on X. Then there exists a measurable G-invariant function $\tilde{f} : X \to \mathbb{R}$ such that $\tilde{f} = f$ almost everywhere on X.*

Proof Replacing f by $\phi \circ f$ for a homeomorphism $\phi : \mathbb{R} \to (0,1)$, we may clearly assume that $f(X) \subseteq (0,1)$.

Fix a probability measure λ on G in the measure class of the Haar measure of G. The subset

$$Q = \{(g,x) \in G \times X \mid f(gx) \neq f(x)\}$$

of $G \times X$ is measurable and, for all $g \in G$,

$$\int_X \chi_Q(g,x)d\mu(x) = \mu(\{x \in X \mid f(gx) \neq f(x)\}) = 0,$$

where χ_Q denotes the characteristic function of Q. Let X_0 be the set of all $x \in X$ for which

$$\int_G \chi_Q(g,x)d\lambda(g) = \lambda(\{g \in G \mid f(gx) \neq f(x)\}) = 0.$$

As G and X are σ-finite measure spaces, Fubini's theorem applies and shows that X_0 is measurable and that $\mu(X \setminus X_0) = 0$.

Let X_1 be the subset of all $x \in X$ for which the mapping $g \mapsto f(gx)$ is constant almost everywhere on G (that is, constant outside a subset of G of measure zero). Clearly, X_1 is G-invariant. Define

$$F : X \to (0,1), \quad F(x) = \int_G f(gx)d\lambda(g)$$

Then, again by Fubini's theorem, F is measurable. Observe that F agrees with f on the set X_0. The set

$$\{(g,x) \in G \times X \mid f(gx) \neq F(x)\}$$

is measurable. As above, Fubini's theorem shows that the set X_2 of all $x \in X$ for which

$$\lambda(\{g \in G \mid f(gx) \neq F(x)\}) = 0,$$

is measurable and that $\mu(X \setminus X_2) = 0$. Now, it is clear that $X_2 = X_1$. Fix any $a \in (0,1)$, and define $\widetilde{f} : X \to (0,1)$ by $\widetilde{f}(x) = F(x)$ if $x \in X_1$ and $\widetilde{f}(x) = a$ if $x \in X \setminus X_1$. Then $\widetilde{f} : X \to (0,1)$ is a measurable G-invariant function and $\widetilde{f} = f$ on X_0. $\square$

The following rephrasis of ergodicity will allow us in §2 to give a representation theoretic formulation of ergodicity when μ is invariant and finite.

1.3 Theorem. *Let G be a locally compact second countable group acting on a σ-finite measure space (X,μ). Then the following are equivalent:*
(i) *The action of G is ergodic;*
(ii) *If $f : X \to \mathbb{R}$ is measurable and essentially G-invariant, then f is constant almost everywhere.*

Proof Let A be a G-invariant measurable subset of X. Then χ_A, the characteristic function of A, is G-invariant. Hence, (ii) implies (i).

To show that (i) implies (ii), assume that G acts ergodically on X. Let f be a measurable and essentially G-invariant function on G. By the previous lemma, we may assume that f is G-invariant. For $k \in \mathbb{Z}$, and $n \in \mathbb{N}$, define

$$X(k,n) = \left\{ x \in X \mid \frac{k}{2^n} \le f(x) < \frac{k+1}{2^n} \right\} .$$

Clearly, $X(k,n)$ is measurable and G-invariant. Hence, by ergodicity,

$$\mu\left(X(k,n)\right) = 0 \quad \text{or} \quad \mu\left(X \setminus X(k,n)\right) = 0.$$

Fix $n \in \mathbb{N}$. Then

$$X = \biguplus_{k \in \mathbb{Z}} X\left(k,n\right) \quad \text{(disjoint union)}.$$

Hence, there exists $k_n \in \mathbb{Z}$ so that

$$\mu(X \setminus X(k_n, n)) = 0.$$

Define

$$Y := \bigcap_{n=1}^{\infty} X\left(k_n\,,\, n\right) = X \setminus \left(\bigcup_{n=1}^{\infty} X \setminus X\left(k_n\,,\, n\right) \right) .$$

Then $\mu(X \setminus Y) = 0$, and f is constant on Y.   □

1.4 Examples.

(i) **Classical mechanics.**

Examples of measure preserving actions arise in classical mechanics (see [AA]). The motion of k particles in the state space Ω, governed by Hamilton's equations

$$\frac{dq_i}{dt} = \frac{\partial H}{\partial p_i}, \quad \frac{dp_i}{dt} = -\frac{\partial H}{\partial q_i} \quad (1 \le i \le k)$$

defines a measure preserving action

$$\mathbb{R} \times \Omega \to \Omega, \quad (t,\omega) \mapsto \varphi_t(\omega)$$

of $\mathbb{R}$ on Ω, where Ω, a subset of $\mathbb{R}^{6k}$, with coordinates $(p,q) \in \mathbb{R}^{6k} = \mathbb{R}^{3k} \times \mathbb{R}^{3k}$, is equipped with Liouville measure $dp\,dq$ and $\varphi_t(\omega)$ denotes the state of the system at time t starting from $\omega \in \Omega$ at time 0. (The invariance of $dp\,dq$ follows from Liouville's theorem; see Example 1.9 (i) below.)

An interesting question is the so-called *ergodic hypothesis of Bolzmann*:
let Ω_E be a level surface for the energy and let $f : \Omega_E \to \mathbb{R}$ be
a measurable function. Is the time average of f equal to its space
average, that is, does

$$\lim_{T \to \infty} \frac{1}{T} \int_0^T f\left(\varphi_t\left(\omega\right)\right) dt = \int_{\Omega_E} f \, dp \, dq$$

hold for $\omega \in \Omega_E$?

In particular, does, for $A \subseteq \Omega_E$, the limit $\lim_{T \to \infty} \frac{r(T)}{T}$ agree with
the volume of A, where $r(T)$ is the measure of the set

$$\{s \mid 0 \le s \le T \text{ such that } \varphi_s\left(\omega\right) \in A\}$$

(ii) Let $X = \mathbb{T} = \{z \in \mathbb{C} \mid |z| = 1\}$ be the one-dimensional torus with
normalized Lebesgue measure μ. Take $\alpha \in \mathbb{R}$ and define

$$\varphi : \mathbb{T} \to \mathbb{T} , \quad \varphi(z) = e^{2\pi i \alpha} z .$$

Clearly, φ preserves μ. Hence φ defines an action of the integers $\mathbb{Z}$
on $\mathbb{T}$ via

$$n \cdot z = \varphi^n\left(z\right) , \quad \forall n \in \mathbb{Z}.$$

Two cases may occur:
• If $\alpha \in \mathbb{Q}$, then $\varphi^N = \text{id}_{\mathbb{T}}$ for some $N \in \mathbb{N}$. It is clear that φ is
not ergodic.
• If $\alpha \notin \mathbb{Q}$, then φ is ergodic.
Indeed, let A be a measurable and φ-invariant subset of $\mathbb{T}$, and let
χ_A be its characteristic function. Let a_n be the n-th Fourier coefficient of χ_A. The n-th Fourier coefficient of the characteristic function
of

$$e^{-2\pi i \alpha} A = \varphi^{-1}(A)$$

is $a_n \, e^{2\pi i n \alpha}$. Hence, by invariance of A,

$$a_n \, e^{2\pi i n \alpha} = a_n, \quad \forall n \in \mathbb{Z}.$$

Since α is irrational, this implies

$$a_n = 0, \quad \forall n \in \mathbb{Z}, n \neq 0.$$

As χ_A coincides, as L^2-function, with its Fourier series

$$z \mapsto \sum_{n=-\infty}^{\infty} a_n z^n,$$

this shows that χ_A is constant almost everywhere. Hence, $\mu\left(A\right) = 0$
or $\mu(A) = 1$.

(iii) The group $G = \mathrm{SL}(n, \mathbb{Z})$ of all $n \times n$ matrices with integer entries and with determinant 1 acts on $\mathbb{R}^n$ in a natural way, preserving the lattice $\mathbb{Z}^n$. Hence, G acts (as a group of automorphisms) on the n-dimensional torus

$$\mathbb{T}^n = \mathbb{R}^n / \mathbb{Z}^n.$$

Moreover, G preserves the Lebesgue measure μ on $\mathbb{T}^n$. We claim that each matrix γ in $\mathrm{SL}(n, \mathbb{Z})$ which has no root of unity as eigenvalue acts ergodically on $\mathbb{T}^n$. As in the above example, the proof will use Fourier analysis. Indeed, let A be a measurable γ-invariant subset of $\mathbb{T}^n$, and, for $k = (k_1, \ldots, k_n) \in \mathbb{Z}^n$, let

$$a_k = \int_{\mathbb{T}^n} \chi_A(z) e^{-2\pi i z.k} dz$$

denote the corresponding Fourier coefficient of χ_A, where

$$z.k := z_1 k_1 + \cdots + z_n k_n$$

for $z = (z_1, \ldots, z_n) \in \mathbb{T}^n$ and dz is the normalized Lebesgue measure on $\mathbb{T}^n$. The Fourier coefficient of $\chi_{\gamma A}$ corresponding to k is $a_{\gamma^t k}$. Hence, by invariance of A,

$$a_k = a_{\gamma^t k}, \quad \forall k \in \mathbb{Z}^n.$$

But, for any $k \in \mathbb{Z}^n$, $k \neq 0$, the set $\{\gamma^n k \mid n \in \mathbb{Z}\}$ is infinite, since γ has no root of unity as eigenvalue. As $\sum_{k \in \mathbb{Z}^n} |a_k|^2$ is finite, a_k has to be zero for $k \in \mathbb{Z}^n \setminus \{0\}$. As above, this shows that χ_A is constant.

Example (iii) generalizes as follows to an action of a group G on a compact abelian group X. We recall a few facts about the duality theory of a compact abelian group X. (For more details, see [HeR], Chap. VI.) The *dual group* $\widehat{X}$ of X is the group of all continuous unitary characters $\sigma : X \to \mathbb{T}$. It is discrete for the topology of uniform convergence.

Let X be equipped with its normalized Haar measure dx. The Fourier coefficients of a function f in $L^1(X)$ are the complex numbers

$$\hat{f}(\sigma) := \int_X f(x) \overline{\sigma(x)} dx, \quad \forall \sigma \in \widehat{X}.$$

The Fourier transform

$$L^2(X) \to \ell^2(\widehat{X}), \quad f \mapsto \hat{f} = (\hat{f}(\sigma))_{\sigma \in \widehat{X}}$$

is a Hilbert space isometry (this is the Plancherel Theorem).

If a group G acts by automorphisms on X, then the dual action of G

on $\widehat{X}$ is of course given by

$$g\sigma(x) = \sigma(g^{-1}x), \quad \forall g \in G, \sigma \in \widehat{X}, x \in X.$$

Observe that the (normalized) Haar measure μ on X is automatically preserved by G. Indeed, let φ be an automorphism of X, and let $\varphi_*(\mu)$ be the image of μ under φ. Then, by unicity of Haar measure, $\varphi_*(\mu)$ is a multiple of μ. On the other hand, $\varphi_*(\mu)(1) = \mu(1)$. This shows that $\varphi_*(\mu) = \mu$.

1.5 Proposition. *Let G be a group acting as a group of automorphisms of a compact abelian group X. Let X be equipped with its normalized Haar measure. Then the action of G on X is ergodic if and only if, except the trivial character $\{1_X\}$, all the G-orbits for the dual action on $\widehat{X}$ are infinite.*

Proof The proof follows *mutatis mutandis* the one given in example (iii) above.

Indeed, let A be a G-invariant measurable subset of X. For $f \in L^2(X)$ and $g \in G$, let $T_g f$ be the function on X defined by

$$T_g f(x) = f(gx), \quad \forall x \in X.$$

The Fourier coefficient of $T_g f$ at $\sigma \in \widehat{X}$ is

$$\widehat{T_g f}(\sigma) = \hat{f}(g\sigma).$$

Take now $f = \chi_A$, the characteristic function of A. Since A is G-invariant, the above shows that the Fourier transform $\hat{f}$ of f is constant on the G-orbits in $\widehat{X}$. Observe that $\hat{f} \in \ell^2(\widehat{X})$ and that f is constant if and only if $\hat{f}(\sigma) = 0$ for all $\sigma \in \widehat{X}$, $\sigma \neq 1_X$. This proves the claim. $\square$

An interesting feature of the above examples is the use of the powerful Fourier analysis methods in order to establish ergodicity. This is not an accidental fact. Indeed, we shall see in the next section how ergodicity fits in the more general framework of unitary group representations.

1.6 Exercise. Let $Y = \{0, 1, \ldots, k-1\}$, $k \in \mathbb{N}$, with the probability measure p,

$$p(i) := 1/k \quad \forall 0 \le i \le k.$$

Let $X = Y^{\mathbb{Z}}$ be the product space $\prod_{n \in \mathbb{Z}} Y$, with the product measure μ. Let

$$\varphi : X \to X$$

be the *Bernoulli-shift* defined by

$$\varphi\big((x_n)_{n\in\mathbb{Z}}\big) = (x_{n+1})_{n\in\mathbb{Z}}.$$

Then φ preserves μ. Show that φ is ergodic .

Hint: Write $Y = \mathbb{Z}/k\mathbb{Z}$, and view X as a compact abelian group.
Then μ is the normalized Haar-measure on Y and φ is an automorphism
of X. More general shifts are obtained when p is an arbitrary probability
measure on Y; concerning their egodicity, see [Wal], Theorem 1.12.

An interesting phenomenon in the case of a finite invariant measure is
recurrence.

1.7 Theorem (Poincaré's Recurrence Theorem). *Let (X,μ) be a
finite measure space, and let $\varphi : X \to X$ be a measure preserving mapping.
Let A be a measurable subset of X. Then almost every point $x \in A$ is
infinitely recurrent with respect to A, that is, the set*

$$\{n \in \mathbb{N}|\ \varphi^n x \in A\}$$

is infinite.

Proof Fix $m \in \mathbb{N}$, and set

$$B_m := \{x \in A|\ \varphi^n x \notin A,\ \forall n \ge m\}.$$

Since $B_m = A \setminus \bigcup_{j \ge m} \varphi^{-j}(A)$, the set B_m is measurable. It is clear that
$B_m \cap \varphi^{-j}(B_m) = \emptyset$ for each $j \ge m$. Hence,

$$\varphi^{-j}(B_m) \cap \varphi^{-k}(B_m) = \emptyset$$

for $j - k \ge m$. Thus $\{\varphi^{-jm}(B_m)\}_{j\in\mathbb{N}_0}$ is a disjoint sequence of measurable
sets. By invariance of μ, all these sets have the same measure. Since $\mu(X)$
is finite, this implies that $\mu(B_m) = 0$. $\qquad\qquad\square$

Very often, the measure space X comes with a (natural) topology. In
that case, we have the following topological version of Poincaré's Recurrence Theorem.

**1.8 Corollary (Poincaré's Recurrence Theorem – Topological Ver-
sion).** *Let (X,d) be a separable metric space, μ a finite Borel measure
on X and $\varphi:X \to X$ a measure preserving mapping. Then μ-almost
every point $x \in X$ is recurrent, that is, there is a sequence of integers
$n_1 < n_2 < \cdots$ such that*

$$\lim_{k\to\infty} \varphi^{n_k} x = x.$$

Proof Let $\{U_n\}_n$ be a countable basis for the topology of X, and let $\widetilde{X}$ be the collection of all recurrent points in X. Then

$$X \setminus \widetilde{X} = \bigcup_n Y_n,$$

where Y_n is the set of all points $x \in U_n$ which are not infinitely recurrent with respect to U_n. By Poincaré's recurrence theorem, $\mu(Y_n) = 0$ and, hence, $\mu(X \setminus \widetilde{X}) = 0$. $\qquad\qquad\square$

1.9 Examples.
(i) **Liouville's Theorem.**
Let $U \subseteq \mathbb{R}^n$ be a bounded open set and let $f \in C^\infty(U, \mathbb{R}^n)$. Consider the autonomous ordinary differential equation

$$\dot{x} = f(x). \qquad\qquad (*)$$

By the standard existence and uniqueness theorem of ordinary differential equations, there exists, for any $p \in U$, a unique solution $\varphi_t(p)$ of $(*)$, with $\varphi_0(p) = p$. For simplicity, we assume that $\varphi_t(p)$ is defined for all $t \in \mathbb{R}$. We then have a one-parameter family of diffeomorphisms $\{\varphi_t\}_{t \in \mathbb{R}}$ which satisfies $\varphi_{t+s} = \varphi_t \circ \varphi_s$.
We claim that the Lebesgue measure λ on U is invariant under the flow $\{\varphi_t\}_t$ if and only if $\mathrm{div} f = 0$, where $\mathrm{div} f = \partial_{x_1} f_1 + \cdots + \partial_{x_n} f_n$ is the divergence of the vector field $f = (f_1, \ldots, f_n)$.
Indeed, for an arbitrary compact subset A of $\mathbb{R}$ and for $h \in \mathbb{R}$, one has

$$\lambda(\varphi_{t+h}(A)) = \int_{\varphi_{t+h}(A)} dx = \int_{\varphi_t(A)} \left| \det \frac{\partial \varphi_h(x)}{\partial x} \right| dx.$$

A Taylor expansion around x shows that $\frac{\partial \varphi_h(x)}{\partial x} = \mathrm{id} + df(x)h + o(h)$, and hence

$$\det \frac{\partial \varphi_h(x)}{\partial x} = 1 + \mathrm{div} f(x) h + o(h).$$

Observe that $\det \frac{\partial \varphi_h(x)}{\partial x}$ is positive for h small enough. Hence,

$$\frac{d}{dt} \lambda(\varphi_t(A)) = \lim_{h \to 0} \int_{\varphi_t(A)} \frac{\mathrm{div} f(x) h + o(h)}{h} dx = \int_{\varphi_t(A)} \mathrm{div} f(x) dx.$$

Thus, if $\mathrm{div} f = 0$, then $\lambda(\varphi_t(A))$ is constant. By Poincaré's recurrence theorem

$$\liminf_{t \to \infty} d(\varphi_t(p), p) \leq \liminf_{n \to \infty} d(\varphi_n(p), p) = \liminf_{n \to \infty} d(\varphi_1^n(p), p) = 0$$

for almost all $p \in U$. The conclusion is that almost all orbits satisfy a recurrence property known as stability in the sense of Poisson.

(ii) Poincaré 's recurrence theorem and Liouville's theorem have the following paradoxical prediction. Consider a box containing some gas. Assume that at the initial time all gas molecules are concentrated in one half of the box. Then, after some time – in fact, infinitely often – the gas molecules will collect in the original portion of space. The resolution is this paradox is the following. The number of degrees of freedom is very large (under normal conditions, the number of gas particles in $1\,\mathrm{cm}^3$ is 10^{20}). Thus, the probability that the gas is in one half of the box behaves like $\lambda^{10^{20}}$, for some $\lambda < 1$, and the approximate period of the Poincaré recurrence is proportional to the inverse of this probability. It is clearly impossible to observe systems during such huge time intervals.

Topological properties such as compactness imply the existence of recurrent points.

1.10 Theorem (Birkhoff). *Let X be a compact space and $\varphi\colon X \to X$ be a continuous map. Then there is a recurrent point $x \in X$.*

Proof Consider the set $\mathcal{O}$ of all non-empty, closed and φ-invariant subsets of X. The finite intersection property shows that $\mathcal{O}$ is inductively ordered by inclusion. Hence, by Zorn's lemma, $\mathcal{O}$ contains a minimal element Y. Clearly, $Y = \overline{\{\varphi^n y\mid n \in \mathbb{N}\}}$ for all $y \in Y$, showing that every $y \in Y$ is recurrent. $\qquad\qquad\Box$

1.11 Examples.

(i) Let G be a *compact* group, $a \in G$ and $\varphi\colon G \to G$, $g \mapsto ag$. Then *all* points in G are recurrent. Indeed, by the previous theorem, there exists a recurrent point $x_0 \in G$. Let $g \in G$ be arbitrary and let $u = x_0^{-1}g$. If V is a neighbourhood of g, then $V \cdot u^{-1}$ is a neighbourhood of x_0. Hence, $a^n x_0 \in V \cdot u^{-1}$, that is, $a^n g \in V$ for some $n \in \mathbb{N}$.

(ii) Here is an application to a problem in diophantine approximation:

For any $\alpha \in \mathbb{R}$ and for any $\varepsilon > 0$, there exist integers n, m such that

$$|\alpha n^2 - m| < \varepsilon.$$

As to the proof, consider the two-dimensional torus

$$\mathbb{T}^2 = \{(\vartheta,\vartheta')\mid \vartheta,\vartheta' \in \mathbb{R}/\mathbb{Z}\}.$$

Define a continous mapping $\varphi\colon \mathbb{T}^2 \to \mathbb{T}^2$ by

$$\varphi(\vartheta,\vartheta') := (\vartheta + [\alpha], \vartheta' + 2\vartheta + [\alpha]),$$

where $[\alpha]$ denotes the image of α in $\mathbb{R}/\mathbb{Z}$. We claim that every point is recurrent for φ. Let $(\vartheta, \vartheta') \in \mathbb{T}^2$, and let $Q(\vartheta, \vartheta')$ be the set of limit points of the sequence

$$\{\varphi^n(\vartheta, \vartheta') | \ n \in \mathbb{N}\}.$$

We have to show that $(\vartheta, \vartheta') \in Q(\vartheta, \vartheta')$. Since φ commutes with translation in the second variable, $Q(\vartheta, \vartheta') = Q(\vartheta, 0)(0, \vartheta')$. Hence, it suffices to show that $(\vartheta, 0) \in Q(\vartheta, 0)$. By (i), ϑ is recurrent with respect to the addition by $[\alpha]$ in $\mathbb{T}$. This, together with the compactness of $\mathbb{T}$, shows that there exists $\vartheta_1 \in \mathbb{T}$ such that $(\vartheta, \vartheta_1) \in Q(\vartheta, 0)$. It follows that $(\vartheta, 2\vartheta_1) \in Q(\vartheta, \vartheta_1)$, and, hence, $(\vartheta, 2\vartheta_1) \in Q(\vartheta, 0)$. Therefore, by induction, $(\vartheta, n\vartheta_1) \in Q(\vartheta, 0)$, for all $n \in \mathbb{N}$. Now, by (i), 0 is in the closure of the set $\{n\vartheta_1 | n \in \mathbb{N}\}$. This proves the claim.

Finally, the orbit of $(0,0)$ is $\varphi^n(0,0) = ([n\alpha], [n^2\alpha])$. In particular, there is a subsequence $\{[n_k^2\alpha]\}_k$ converging to zero in $\mathbb{T}$, and this finishes the proof.

We conclude this section by discussing examples of flows. A *flow* is simply an action of $\mathbb{R}$ on some measure space (X, μ).

1.12 Examples.
(i) The most familiar examples of flows come from autonomous differential equations as discussed in 1.9 (i)

$$\dot{x} = f(x).$$

If the divergence of f vanishes, the Lebesgue measure is invariant under the induced flow.

(ii) Consider $X = \mathbb{T}^n$ with the normalized Lebesgue measure. For $\alpha_1, \ldots, \alpha_n$ in $\mathbb{R}$, define a flow on $\mathbb{T}^n$ by

$$\varphi_t(z_1, \ldots, z_n) = \left(e^{2\pi i \alpha_1 t} z_1, \ldots, e^{2\pi i \alpha_n t} z_n\right)$$

for any $t \in \mathbb{R}$. The flow $\{\varphi_t | \ t \in \mathbb{R}\}$ is ergodic if and only if $\alpha_1, \ldots, \alpha_n$ are linearly independent over $\mathbb{Q}$, that is, if $k_1 \alpha_1 + \cdots + k_n \alpha_n = 0$ for some $k_1, \ldots, k_n \in \mathbb{Z}$, then $k_1 = \cdots = k_n = 0$. This can be shown by a Fourier analysis argument as in Examples 1.4 (ii). Indeed, let $f \in L^2(\mathbb{T}^n)$ be invariant under all φ_t, $t \in \mathbb{R}$. Then, denoting by a_k its Fourier coefficients and with $\alpha = (\alpha_1, \ldots, \alpha_n)$,

$$a_k = a_k \, e^{2\pi i t k \cdot \alpha} \, , \quad \forall k \in \mathbb{Z}^n, \ t \in \mathbb{R}$$

and, hence,

$$e^{2\pi i t k \cdot \alpha} = 1 \, , \quad \forall t \in \mathbb{R}$$

whenever $a_k \neq 0$. So, $a_k \neq 0$ implies $k \cdot \alpha = 0$, showing that f is constant almost everywhere if and only if $\alpha_1, \alpha_2, \ldots, \alpha_n$ are linearly independent over $\mathbb{Q}$.

(iii) **Kronecker's Theorem.**

Let $\{\varphi_t\}_{t \in \mathbb{R}}$ be the above flow on $\mathbb{T}^n$, with $\alpha_1, \alpha_2, \ldots, \alpha_n$ linearly independent over $\mathbb{Q}$. It is remarkable that *all* the orbits of this flow are dense in $\mathbb{T}^n$. The flow is then said to be *minimal*.

Indeed, let $z \in \mathbb{T}^n$. Since

$$\{\varphi_t(z)|\ t \in \mathbb{R}\} = z\,\{\varphi_t(1)|\ t \in \mathbb{R}\}$$

(where $1 = (1, \ldots, 1)$ is the group unit of $\mathbb{T}^n$), it suffices to show that the set $\{\varphi_t(1)|\ t \in \mathbb{R}\}$ is dense in $\mathbb{T}^n$. Now

$$\{\varphi_t(1)\,|\,t \in \mathbb{R}\} = \left\{ \left(e^{2\pi i t \alpha_1}, \ldots, e^{2\pi i t \alpha_n} \right) \,|\, t \in \mathbb{R}\right\}$$

is a *subgroup* of $\mathbb{T}^n$. The closure H of this subgroup is again a subgroup of $\mathbb{T}^n$. Suppose, by contradiction, that $H \neq \mathbb{T}^n$. Then the quotient $\mathbb{T}^n/H$ is a non-trivial abelian locally compact group and, hence, has non-trivial characters. Therefore, there exists a character χ on $\mathbb{T}^n$ with $\chi \neq 1$ and such that $\chi|_H = 1$. Let $k = (k_1, \ldots, k_n) \in \mathbb{Z}^n \setminus \{0\}$ be such that

$$\chi(z) = z_1^{k_1} \ldots z_n^{k_n}, \quad \forall\, z = (z_1, \ldots, z_n) \in \mathbb{T}^n .$$

Then

$$\chi\left(e^{2\pi i t\, \alpha_1}, \ldots, e^{2\pi i t \alpha_n} \right) = e^{2\pi i t \alpha \cdot k} = 1$$

for all $t \in \mathbb{R}$, which implies $\alpha \cdot k = 0$, a contradiction.

(iv) The previous example may be generalized as follows. Let $\{\varphi_t\}_{t \in \mathbb{R}}$ be the above flow on $\mathbb{T}^n$, where now $\alpha_1, \alpha_2, \ldots, \alpha_n$ are not necessarily linearly independent over $\mathbb{Q}$. Let r be the dimension over $\mathbb{Q}$ of the $\mathbb{Q}$-linear span of $\alpha_1, \alpha_2, \ldots, \alpha_n$. Then there exists a closed subgroup H of $\mathbb{T}^n$ isomorphic to $\mathbb{T}^r$ such that, for any $z \in \mathbb{T}^n$, the closure of the orbit $\{\varphi_t(z)\}_{t \in \mathbb{R}}$ coincides with Hz. We shall discuss in Chap. VI – in relation with Raghunathan's conjecture – a far-reaching generalization of this fact.

As to the proof, let H be the closure of the subgroup $\{\varphi_t(1)\}_{t \in \mathbb{R}}$. The set $H^\perp$ of all characters annihilating H may be identified (as above) with the subgroup Δ of all $k \in \mathbb{Z}^n$ with $\alpha \cdot k = 0$. Observe that Δ is a *complete* subgroup of $\mathbb{Z}^n$, that is, $\Delta = V \cap \mathbb{Z}^n$ for a linear subspace V (of dimension $n - r$) of $\mathbb{Q}^n$. Hence, there exists $g \in \mathrm{SL}(n, \mathbb{Z})$ such that $\Delta = g\mathbb{Z}^{n-r}$, where $\mathbb{Z}^{n-r}$ is identified with the subgroup $\mathbb{Z}^{n-r} \times \{0\}$ of $\mathbb{Z}^n$. (Equivalently, a $\mathbb{Z}$-basis of Δ can be extended to a basis of $\mathbb{Z}^n$; see Chap. VI, Exercise 3.5.)

By standard duality theory,

$$H = (H^\perp)^\perp = \Delta^\perp$$

and, hence (recall that $\mathrm{SL}(n, \mathbb{Z})$ acts on $\mathbb{T}^n$),

$$g^t H = (\mathbb{Z}^{n-r})^\perp = \{1\} \times \mathbb{T}^r.$$

Thus, $H \cong \mathbb{T}^r$.

1.13 Exercise. For $\alpha_1, \ldots, \alpha_n \in \mathbb{R}$, define

$$\varphi : \mathbb{T}^n \to \mathbb{T}^n, \, \varphi(z_1, \ldots, z_n) = \left(e^{2\pi i \alpha_1} z_1, \ldots, e^{2\pi i \alpha_n} z_n\right).$$

Show that the action of $\mathbb{Z}$ on $\mathbb{T}^n$ is ergodic if and only if $\alpha_1, \ldots, \alpha_n, 1$ are linearly independent over $\mathbb{Q}$. Moreover, in that case, all the orbits are dense in $\mathbb{T}^n$.

1.14 Exercise (Weyl's Equidistribution Theorem). Let $\alpha_1, \ldots, \alpha_n \in \mathbb{R}$ be such that $\alpha_1, \ldots \alpha_n, 1$ are linear independent over $\mathbb{Q}$. Then for any $f \in C(\mathbb{T}^n)$ and for *any* $(z_1, \ldots, z_n) \in \mathbb{T}^n$,

$$\lim_{N \to \infty} \frac{1}{N} \sum_{k=0}^{N-1} f\left(e^{2\pi i k \alpha_1} z_1, \ldots, e^{2\pi i k \alpha_n} z_n\right) = \int_{\mathbb{T}^n} f \, d\mu,$$

where μ is the normalized Haar measure on $\mathbb{T}^n$.

Hint: Prove the formula first for trigonometric polynomials.

§2 Ergodic Theory and Unitary Representations

In this section, we shall consider a measure preserving action of a locally compact group G on a probability space (X, μ). For a measurable function $f : X \to \mathbb{C}$ and for $g \in G$, define

$$(T_g f)(x) = f(g^{-1} x), \quad x \in X.$$

Since G is measure-preserving, it is clear that

$$T_g : L^2(X) \to L^2(X)$$

is a unitary operator on $L^2(X) = L^2(X, \mu)$ and that the mapping

$$G \to L^2(X), \quad g \mapsto T_g f$$

is continuous for all $f \in L^2(X)$. Hence, $T : g \mapsto T_g$ is a *unitary representation* of G on $L^2(X)$. Recall that a unitary representation of G on a Hilbert space $\mathcal{H}$ is a homomorphism π from G into the group $\mathcal{U}(\mathcal{H})$

of all unitary operators on $\mathcal{H}$, which is continuous in the strong operator topology, that is, the mapping

$$G \to \mathcal{H}, \quad g \mapsto \pi(g)\xi$$

is continuous for all $\xi \in \mathcal{H}$.

The following theorem, first observed by Koopman, gives a characterization of ergodicity in terms of the above unitary representation T. It is often a useful tool for proving ergodicity.

2.1 Theorem. *A measure-preserving action of G on a probability space (X, μ) is ergodic if and only if the space*

$$\{f \in L^2(X) \,|\, T_g f = f, \ \forall g \in G\}$$

of the invariant vectors consists only of the constants.

Proof If the action is ergodic, then, by Theorem 1.3, the constants are the only essentially G-invariant measurable functions on X. If the action is not ergodic, then there exists an invariant set A with measure

$$0 < \mu(A) < 1.$$

The characteristic function of A is then an invariant non-constant L^2-function on X. $\qquad\Box$

The following is a rephrasing of the above theorem.

2.2 Theorem. *Define*

$$L_0^2(X) := (\mathbb{C} \cdot 1)^{\perp} = \left\{ f \in L^2(X) \,|\, \int_X f d\mu = 0 \right\},$$

the (G-invariant) space of functions with zero mean. Let T^0 be the unitary representation of G obtained by restricting the T_g's to $L_0^2(X)$. Then the action of G is ergodic if and only if T^0 does not contain the trivial representation 1_G of G, that is, there is no non-zero subspace on which T^0 is trivial.

The finiteness of the invariant measure is crucial in the above theorem. For example, the group of integers $\mathbb{Z}$ acts on the reals via translations. This action is not ergodic (consider, for instance, the invariant set $\bigcup_{n \in \mathbb{Z}} [n, n + \frac{1}{2}]$). On the other hand, let f be a function in $L^2(\mathbb{R})$ which is invariant by translations with integers. The restrictions of f to the cosets $x + \mathbb{Z}$ are square summable and hence 0 for almost every $x \in \mathbb{R}$. Therefore, $f = 0$.

Some Classical Ergodic Theorems

Let (X, μ) be a probability space and $\varphi : X \to X$ a measure-preserving bijection. Thus, φ defines an action of $\mathbb{Z}$ on X. Furthermore, let $f : X \to \mathbb{C}$ be a measurable function.

Consider the time averages of f defined by

$$\frac{1}{n+1} \sum_{i=0}^{n} f\left(\varphi^i x\right) = \frac{1}{n+1} \sum_{i=0}^{n} (T^{-i}f)(x), \qquad x \in X,$$

where

$$Tf(x) := f(\varphi^{-1}x).$$

The abstract form of Boltzman's ergodic hypothesis (see 1.4 (i)) is the question as to whether the equality

$$\lim_{n\to\infty} \frac{1}{n+1} \sum_{i=0}^{n} f(\varphi^i x) = \int_X f \, d\mu$$

holds, that is, whether time and space averages of f coincide in the long run.

Since we are considering arbitrary measurable functions, the (asymptotic) equality of the averages above may be considered in different senses: pointwise convergence or convergence in the quadratic mean, for example. In the case of ergodicity, as we shall now see, both questions have a positive answer.

2.3 Theorem (von Neumann's Ergodic Theorem). *Assume that $\varphi : X \to X$ is ergodic. Then*

$$\lim_{n\to\infty} \left\| \frac{1}{n+1} \sum_{i=0}^{n} T^{-i} f - \left(\int_X f \, d\mu \right) \cdot 1 \right\|_2 = 0$$

for any $f \in L^2(X)$, where 1 denotes the function constant equal to 1 on X.

This theorem is an immediate consequence of the following.

2.4 Theorem (Abstract Ergodic Theorem). *Let $\mathcal{H}$ be a Hilbert space, and let $T : \mathcal{H} \to \mathcal{H}$ be a unitary operator on $\mathcal{H}$. Let*

$$\mathcal{H}_0 = \{\xi \in \mathcal{H} \mid T\xi = \xi\}$$

be the fixed space of T, and let $P : \mathcal{H} \to \mathcal{H}_0$ be the orthogonal projection *on* $\mathcal{H}_0$. *Then*

$$\lim_{n\to\infty} \left\| \frac{1}{n+1} \sum_{i=0}^{n} T^i \xi - P\xi \right\| = 0, \quad \forall \xi \in \mathcal{H}.$$

Before proving this theorem, we indicate how it implies von Neumann's ergodic theorem. Take $\mathcal{H} := L^2(X)$, and let T be defined as before. By ergodicity,

$$\mathcal{H}_0 = \{f \in L^2(X) | Tf = f\}$$

consists of the constants. Hence,

$$Pf = \left(\int_X f \, d\mu \right) \cdot \mathbf{1}.$$

Proof of the Abstract Ergodic Theorem. Clearly, the theorem is true for $\xi \in \mathcal{H}_0$. It thus suffices to show that

$$\lim_{n\to\infty} \left\| \frac{1}{n+1} \sum_{i=0}^{n} T^i \xi \right\| = 0,$$

for all $\xi \in \mathcal{H}_0^\perp$. One has

$$\mathcal{H}_0 = \{T\xi - \xi \mid \xi \in \mathcal{H}\}^\perp.$$

Indeed, using $T^* = T^{-1}$, we find

$$\langle \eta, T\xi - \xi \rangle = \langle T^{-1}\eta, \xi \rangle - \langle \eta, \xi \rangle = 0, \quad \forall \eta \in \mathcal{H}_0.$$

Conversely, if $\langle \eta, T\xi - \xi \rangle = 0$ for all $\xi \in \mathcal{H}$, then $T^{-1}\eta = \eta$, that is, $T\eta = \eta$.

Therefore, $\mathcal{H}_0^\perp$ is the closure of $\{T\eta - \eta \mid \eta \in \mathcal{H}\}$. So, it suffices to prove the theorem for ξ of the form $\xi = T\eta - \eta$. Now,

$$\left\| \frac{1}{n+1} \sum_{i=0}^{n} T^i (T\eta - \eta) \right\| = \left\| \frac{1}{n+1} \left(T\eta - \eta + T^2\eta - T\eta + \cdots + T^{n+1}\eta - T^n\eta \right) \right.$$

$$= \left\| \frac{1}{n+1}(-\eta + T^{n+1}\eta) \right\| \le \frac{2}{n+1} \|\eta\| ,$$

and, hence,

$$\lim_{n\to\infty} \left\| \frac{1}{n+1} \sum_{i=0}^{n} T^i \xi \right\| = 0.$$

$\square$

The pointwise (almost everywhere) convergence is much harder to prove. This is the famous Birkhoff ergodic theorem. We reproduce below a new, surprisingly elementary proof, due to M. Keane (see [BKS], Theorem 2.2).

2.5 Theorem (Birkhoff's Ergodic Theorem). *Under the same assumptions as in von Neumann's Ergodic Theorem, one has, for every $f \in L^1(X)$,*

$$\lim_{n \to \infty} \frac{1}{n+1} \sum_{i=0}^{n} f(\varphi^i x) = \int_X f \, d\mu$$

for almost every $x \in X$.

Proof Clearly, it suffices to prove the theorem for a measurable, real function f with $0 \le f \le 1$. For $x \in X$, let

$$A_n(x) := \frac{1}{n+1} \sum_{i=0}^{n} f(\varphi^i x).$$

Then

$$\bar{A}(x) := \limsup A_n(x)$$

is a well-defined measurable function. It is enough to show

$$\int_X \bar{A} d\mu \le \int_X f d\mu \tag{1}$$

Indeed, inequality (1) applied to $1 - f$ instead of f implies

$$\int_X \underline{A} d\mu \ge \int_X f d\mu, \tag{2}$$

where $\underline{A}(x) := \liminf A_n(x)$. Now, (1), (2) and $\bar{A} \ge \underline{A}$ imply

$$\int_X (\bar{A} - \underline{A}) \, d\mu = 0,$$

and, hence,

$$\bar{A} = \underline{A}$$

almost everywhere. Since we obviously have

$$\bar{A}(\varphi y) = \bar{A}(y) \quad \forall y \in X,$$

ergodicity forces $\bar{A}$ to be constant almost everywhere and this shows the claim.

To show inequality (1), fix $\varepsilon > 0$ and consider the following well-defined and measurable function

$$\tau : X \to \mathbb{N}_0, \ x \mapsto \min\{n \in \mathbb{N}_0 \mid A_n(x) \ge \bar{A}(x) - \varepsilon\}.$$

Assume first that there exists a constant $M > 0$ such that $\tau(x) \leq M$ for almost all x. Pick $x \in X$ with $\tau(\varphi^k x) \leq M$ for all $k \geq 0$. (This is possible, as, for any integer k, $\tau(\varphi^k x) \leq M$ for almost all $x \in X$.) Then consider the orbit of x until n and group it in the following way:

$$x, \varphi x, \ldots, \varphi^{i_1} x,$$
$$\varphi^{i_1+1} x, \varphi^{i_1+2} x, \ldots, \varphi^{i_2} x, \ldots,$$
$$\varphi^{i_{r-1}+1} x, \varphi^{i_{r-1}+2} x, \ldots, \varphi^{i_r} x,$$
$$\varphi^{i_r+1} x, \varphi^{i_r+2} x, \ldots, \varphi^n x,$$

where the i_k are determined by the condition

$$i_1 = \tau(x), \qquad i_{k+1} - i_k = \tau(\varphi^{i_k+1} x), \qquad 1 \leq k \leq r-1.$$

The remaining piece of the orbit consists of at most M points, by the boundedness of τ (that is, $n - i_r \leq M$). Let

$$S_n(x) = (n+1) A_n(x) = \sum_{i=0}^{n} f(\varphi^i x).$$

Using the invariance of $\bar{A}$, one has

$$S_n(x) =$$
$$= S_{i_1}(x) + S_{i_2-i_1}(\varphi^{i_1+1} x) + \cdots + S_{i_r-i_{r-1}}(\varphi^{i_{r-1}+1} x) + S_{n-i_r}(\varphi^{i_r+1} x)$$
$$\geq S_{i_1}(x) + S_{i_2-i_1}(\varphi^{i_1+1} x) + \cdots + S_{i_r-i_{r-1}}(\varphi^{i_{r-1}+1} x)$$
$$\geq i_1 \cdot (\bar{A}(x) - \varepsilon) + (i_2 - i_1) \cdot (\bar{A}(x) - \varepsilon) + \cdots + (i_r - i_{r-1}) \cdot (\bar{A}(x) - \varepsilon)$$
$$= i_r \cdot (\bar{A}(x) - \varepsilon)$$
$$\geq (n - M) \cdot (\bar{A}(x) - \varepsilon).$$

This inequality holds for any $n \in \mathbb{N}$. Dividing both sides by $n+1$, integrating over X and letting n go to infinity, we obtain

$$\int_X f d\mu \geq \int_X \bar{A} d\mu - \varepsilon,$$

as desired.

If the function τ is not essentially bounded, choose an $M \in \mathbb{N}$, such that

$$\mu(\{x \mid \tau(x) > M\}) < \varepsilon.$$

Such a number M exists due to the finiteness of μ. Let $f' := f + g$, where g is the characteristic function of the set $\{x \mid \tau(x) > M\}$. Let

$S'_n(x) = \sum_{i=0}^n f'(\varphi^i x)$. Observe that for all $x \in X$

$$\min\{n \in \mathbb{N}_0 \,|\, S'_n(x) \geq (n+1)(\bar{A}(x) - \varepsilon)\} \leq M$$

holds. (This is clear if $\tau(x) \leq M$, since $S'_n \geq S_n$, and if $\tau(x) > M$, then this minimum is $n = 0$, since $\bar{A}(x) \leq 1$.)
Now, the analogous decomposition of the orbit yields

$$S'_n(x) \geq (n - M) \cdot (\bar{A}(x) - \varepsilon).$$

As above, we obtain

$$\int_X f' d\mu \geq \int_X \bar{A} \, d\mu - \varepsilon$$

and, since

$$\int_X f' d\mu \leq \int_X f d\mu + \varepsilon,$$

the desired inequality

$$\int_X f \, d\mu \geq \int_X \bar{A} \, d\mu - 2\varepsilon.$$

$\square$

2.6 Remarks and Applications.
(i) Strong Law of Large Numbers.
For $k \in \mathbb{N}$, let $Y = \{0, 1, \ldots, k-1\}$ with probabilities $p_0, p_1, \ldots, p_{k-1}$. Let $X = Y^{\mathbb{Z}}$ with the product measure μ and let $\varphi : X \to X$ be the Bernoulli-shift. Fix $0 \leq j \leq k-1$ and define

$$A = \{(\ldots, x_{-1}, x_0, x_1, \ldots) \in X \,|\, x_0 = j\}.$$

Then $\int_X \chi_A \, d\mu = p_j$ and, by Birkhoff's ergodic theorem, for almost every $\omega \in X$ the sum

$$\frac{1}{n+1} \sum_{i=0}^n \chi_A(\varphi^i \omega),$$

which is the frequency of occurrence of j during the period $[0, n]$, tends to p_j as $n \to \infty$.

(ii) Let (X, μ, φ) be an ergodic sytem, where μ is a φ-invariant probability measure on X. Then, for any measurable sets $A, B \subseteq X$ with $\mu(A) > 0$ and $\mu(B) > 0$, there exist infinitely many $n \in \mathbb{N}$ such that $\mu(\varphi^n A \cap B) > 0$. This says that the orbit $\{\varphi^n A\}_{n \in \mathbb{N}}$ is dense in a measure theoretic sense.

As to the proof, look at

$$A' = \bigcap_{n=0}^{\infty} \bigcup_{i=n}^{\infty} \varphi^i A.$$

It is straightforward that $\varphi A' = A'$ and that

$$\mu \left(\bigcup_{i=0}^{\infty} \varphi^i A \right) = \mu \left(\bigcup_{i=n}^{\infty} \varphi^i A \right)$$

for every $n \in \mathbb{N}$. Hence,

$$\mu(A') = \mu \left(\bigcup_{i=0}^{\infty} \varphi^i A \right) \geq \mu(A) > 0.$$

Hence, by ergodicity, $\mu(A') = 1$ and $\mu(A' \cap B) > 0$. Therefore $\mu(\varphi^n A \cap B) > 0$ for infinitely many n.

Note that, on the other hand, the above property clearly implies the ergodicity of (X, μ, φ).

(iii) An easy corollary of Birkhoff's ergodic theorem is the following strengthening of (ii) above.

The following are equivalent:

(a) (X, μ, φ) is ergodic;

(b) For any measurable subsets $A, B \subseteq X$ one has

$$\lim_{n \to \infty} \frac{1}{n} \sum_{i=0}^{n-1} \mu(\varphi^{-i} A \cap B) = \mu(A)\,\mu(B).$$

Thus, the events $\varphi^{-n} A$ and B become independent in the mean.

To see that (a) implies (b), apply Birkhoff's theorem to $f = \chi_A$ to find

$$\lim_{n \to \infty} \frac{1}{n} \sum_{i=0}^{n-1} \chi_A(\varphi^i x) = \mu(A),$$

and, hence,

$$\lim_{n \to \infty} \frac{1}{n} \sum_{i=0}^{n-1} \chi_{\varphi^{-i} A}(x) \chi_B(x) = \mu(A)\chi_B(x).$$

for almost all $x \in X$. By Lebesgue's theorem of majorized convergence, this shows that

$$\lim_{n \to \infty} \frac{1}{n} \sum_{i=0}^{n-1} \mu(\varphi^{-i} A \cap B) = \mu(A)\,\mu(B).$$

To see that (b) implies (a), let $A \subseteq X$ be measurable and invariant. Setting $B = A$ in (b) we find

$$\mu(A) = \frac{1}{n} \sum_{i=0}^{n-1} \mu(\varphi^{-i} A \cap A) \to \mu(A)\,\mu(A) = \mu(A)^2.$$

Hence, $\mu(A) = 0$ or $\mu(A) = 1$. The property (b) is a weak form of mixing, a notion to be discussed now.

Strong Mixing

Strong mixing is another invariant of measure-preserving group actions.

2.7 Definition. The dynamical system (X, μ, φ), where μ is a φ-invariant probability measure, is *strongly mixing* if, for any measurable sets $A, B \subseteq X$, one has

$$\lim_{n \to \infty} \mu(\varphi^{-n} A \cap B) = \mu(A)\,\mu(B).$$

Strong mixing implies ergodicity (by Remark 2.6 (iii) above). Strong mixing means that the events $\varphi^{-n} A$ and B become independent for n large. This corresponds to the idea of "mixing". To look at a concrete example, consider a bar-mixer putting in a shaker 10 cl vermouth and 90 cl vodka. Then after shaking for a long time, every portion of this drink will have the ratio 1:9 in the mean, if the shaking is ergodic, and it will surely have the correct ratio, if the shaking is strongly mixing (as one might hope).

2.8 Examples.

(i) Let $\alpha \in \mathbb{R} \setminus \mathbb{Q}$ and $\varphi : \mathbb{T} \to \mathbb{T}$, $\varphi(z) := e^{2\pi i \alpha} z$. Then φ is *not* strongly mixing (although it is ergodic by 1.4 (ii)). Indeed, it is clear that, if $A, B \subseteq \mathbb{T}$ are two disjoint and sufficiently small intervals, then $\varphi^n A \cap B = \emptyset$ for infinitely many $n \in \mathbb{N}$.

(ii) The Bernoulli-shift (see Exercise 1.6)

$$\varphi : X \to X, \; \varphi((x_n)_{n \in \mathbb{Z}}) = (x_{n+1})_{n \in \mathbb{Z}} \qquad (X = \{0, 1, \ldots, k-1\}^{\mathbb{Z}})$$

is strongly mixing.
In fact, let A, B be intersections of finitely many sets of the form

$$\{(x_n)_{n \in \mathbb{Z}} \mid x_i = j\}.$$

Clearly, for n large enough,

$$\mu(\varphi^{-n} A \cap B) = \mu(\varphi^{-n} A) \, \mu(B)$$
$$= \mu(A) \, \mu(B) \, .$$

Such sets A, B generate the σ-algebra of the measurable sets in X. Thus the same is true for all measurable subsets A, B. (For more details, see [Wl], Theorem 1.30.)

(iii) (**"Baker transformation"**)
Let $X = [0, 1] \times [0, 1]$ the square with Lebesgue measure, and consider $\varphi : X \to X$ defined by

$$\varphi(x, y) = \begin{cases} \left(2x, \dfrac{y}{2}\right), \, 0 \leq x \leq 1/2 \\[2mm] \left(2x - 1, \dfrac{y+1}{2}\right), \, \dfrac{1}{2} < x \leq 1 \end{cases}$$

Clearly, φ preserves the Lebesgue measure.
Let D be the set (of measure zero) of the dyadic rationals in $[0, 1]$. Now write $x, y \in [0, 1] \setminus D$ in dyadic expansion $x = .x_1 x_2 \ldots$, $y = .y_1 y_2 \ldots$ and the point $(x, y) \in [0, 1]^2$ as

$$(x, y) = (\ldots, y_3, y_2, y_1, x_1, x_2, \ldots).$$

Then

$$\varphi(x, y) = (\ldots, y_2, y_1, x_1, x_2, x_3, \ldots).$$

Thus, up to a set of measure zero, the baker transformation is a realization of the Bernoulli-shift with $Y = \{0, 1\}$.

Recall that the ergodicity of the measure-preserving system on a probability space (X, μ) was described in terms of the associated unitary operator

$$Tf(x) = f(\varphi^{-1} x), \qquad x \in X$$

on $L^2(X)$.

2.9 Theorem. *Let φ be measure-preserving on the probability space (X, μ). The following are equivalent:*

(i) *The system* (X, μ, φ) *is strongly mixing;*

(ii) *For any* $f, g \in L^2(X)$, *one has*

$$\lim_{n \to \infty} \langle T^n f , g \rangle = \langle f, 1 \rangle \langle 1, g \rangle,$$

where $\langle \cdot, \cdot \rangle$ *denotes the inner product on* $L^2(X)$;

(iii) *For any* $f, g \in L_0^2(X) = (\mathbb{C}1)^\perp$, *one has*

$$\lim_{n \to \infty} \langle T^n f , g \rangle = 0.$$

Proof Suppose (i) holds. Then, by definition, the formula (ii) is true for the characteristic functions $f = \chi_A$, $g = \chi_B$ of measurable sets A, B. Now, by density of the simple functions, the formula (ii) holds for arbitrary square integrable functions.

The equivalence of (ii) and (iii) is trivial. Choosing f, g as characteristic functions shows that (ii) implies (i). $\qquad\square$

The following is a natural generalization of strong mixing to group actions. Recall that a function $f : X \to \mathbb{C}$ on a locally compact space X *vanishes at infinity* if the set $\{x \in X \,|\, |f(x)| \geq \varepsilon\}$ is compact, for any $\varepsilon > 0$. We denote by $C_0(X)$ the space of all continuous functions on X that vanish at infinity. If X is compact, $C_0(X)$ is the space of all continuous functions on X and is denoted simply by $C(X)$.

2.10 Definition. The action of the group G on (X, μ) is *strongly mixing* if all the matrix coefficients

$$g \mapsto \langle T_g \xi , \eta \rangle, \quad \xi, \eta \in L_0^2(X)$$

vanish at infinity (that is, belong to $C_0(G)$).

2.11 Remarks and Examples.

(i) We say that a sequence (or a net) $\{x_i\}_i$ in a locally compact space X *diverges to infinity* or *tends to infinity* and write $x_i \to \infty$ if $\{x_i\}_i$ has no limit point in X. A function $f : X \to \mathbb{C}$ vanishes at infinity if and only if $\lim_i f(x_i) = 0$ for every sequence $\{x_i\}_i$ in X with $x_i \to \infty$.

Thus, the action of G on (X, μ) is strongly mixing if and only if, for all sequences $\{g_i\}_i$ in G with $g_i \to \infty$, one has

$$\lim_i \int_X \xi(g_i x) \eta(x) d\mu(x) = 0, \quad \forall \xi, \eta \in L_0^2(X)$$

or, equivalently,

$$\lim_i \int_X \xi(g_i x)\eta(x)d\mu(x) = \int_X \xi d\mu \int_X \eta d\mu, \quad \forall \xi, \eta \in L^2(X).$$

(ii) Note that the notion of strong mixing makes sense only for a *non-compact* group G. Notice also that strong mixing implies ergodicity. Indeed, if $f \in L_0^2(X)$ is G-invariant, then

$$\langle f, f \rangle = \langle T_g f, f \rangle = \lim_{g \to \infty} \langle T_g f, f \rangle = 0.$$

Hence, $f = 0$ and the action is ergodic, by Theorem 2.2.

(iii) Strong mixing is inherited by closed *non-compact* subgroups. That means that, if H is a closed non-compact subgroup of G and if the action of G on X is strongly mixing, then the (induced) action of H on X is also strongly mixing. This property will be useful when we consider actions of semisimple Lie groups in Chap. III.

In contrast, ergodicity is *not* inherited by closed subgroups. For example, the reals act transitively on the circle $\mathbb{T}$ via $\alpha \cdot z := e^{2\pi i \alpha}z$ (for $\alpha \in \mathbb{R}$ and $z \in \mathbb{T}$), whereas the restriction of this action to the integers is trivial.

(iv) The (natural) action of $G = \mathrm{SL}(n, \mathbb{Z})$ on the n-dimensional torus $\mathbb{T}^n$ is *not* strongly mixing for $n \geq 2$ (although it is ergodic as already seen in 1.4 (iii)). To show this, take for instance

$$H = \left\{ \begin{pmatrix} 1 & m \\ 0 & 1 \end{pmatrix} \mid m \in \mathbb{Z} \right\}.$$

The action of H on $\mathbb{T}^2$ is not even ergodic since it fixes the character

$$\chi : \mathbb{T}^2 \to \mathbb{T}, \quad \chi(z_1, z_2) = z_2.$$

This easily generalizes to any $n \geq 2$.

Weak Mixing

We discuss now a weak version of mixing. Ergodicity of the system (X, μ, φ) for an invariant probability measure μ means that 1 is a simple eigenvalue of the associated unitary operator

$$T : L^2(X) \to L^2(X), \quad Tf = f \circ \varphi^{-1}.$$

Weak mixing is defined by means of another spectral property of T as follows.

2.12 Definition. The system (X, μ, φ), where μ is an invariant probability measure, is *weakly mixing* if 1 is a simple eigenvalue of T and if there are no other eigenvalues of T.

2.13 Remarks.

(i) Strong mixing implies weak mixing. Indeed, (X, μ, φ) is strongly mixing if and only if $\langle T^n f, g \rangle \to 0$ as $n \to \infty$ for all $f, g \in L_0^2(X)$.
If the system were not weakly mixing, there would exist a non-zero eigenfunction f with eigenvalue $\lambda \in \mathbb{T}$ which is orthogonal to the constants (since T is unitary, in particular normal). Hence,

$$\langle T^n f, f \rangle = \lambda^n \|f\|^2 \to 0 \quad \text{as} \quad n \to \infty,$$

a contradiction to $|\lambda| = 1$.

(ii) The ergodic system

$$X = \mathbb{T}, \; \varphi(z) = e^{2\pi i \alpha} z \quad (\alpha \in \mathbb{R} \setminus \mathbb{Q})$$

is not weakly mixing, since any character $\chi(z) = z^k \, (k \in \mathbb{Z})$ is an eigenfunction of T with eigenvalue $e^{2\pi i k \alpha}$.

(iii) The following is an equivalent definition of weak mixing more related to the definition of strong mixing (see also the characterization of the ergodicity in 2.6 (iii)).
The following are equivalent:
(i) The system (X, μ, φ) is weakly mixing;
(ii) For any measurable subsets A, B of X,

$$\lim_{n \to \infty} \frac{1}{n} \sum_{i=0}^{n-1} |\mu(\varphi^{-i} A \cap B) - \mu(A)\,\mu(B)| = 0.$$

(For the proof, see [Wl], 1.21.)

The definition of weak mixing systems generalizes as follows to actions of arbitrary groups.

2.14 Definition. Let G be a locally compact group acting on a probability space (X, μ). Then the system (X, μ, G) is called *weakly mixing* if the associated unitary representation T^0 on $L_0^2(X) = (\mathbb{C} \cdot 1)^{\perp}$ contains no finite-dimensional subrepresentation, that is, if there is no G-invariant, non zero finite-dimensional subspace of $L_0^2(X)$.

2.15 Remarks.

(i) In the case of an action of $\mathbb{Z}$ defined by a transformation φ, this definition agrees with the one given before. Indeed, if the system (X, μ, φ) is not weakly mixing, there is a non-zero finite-dimensional T-invariant subspace in $L_0^2(X)$, spanned by an eigenvector. On the other hand, suppose that there exists a non-zero finite-dimensional T-invariant subspace V in $L_0^2(X)$. The restriction $T\big|_V$ is a unitary transformation of V. Hence, $T|_V$ has an eigenvector, and the system (X, μ, φ) is not weakly mixing.

(ii) An action of a compact group is never weakly mixing. Indeed, by the Peter–Weyl theorem (see [BrD], Chap. III), the representation T^0 decomposes into a sum of finite-dimensional ones. Systems with this property are said to have *discrete spectrum*.

(iii) A strongly mixing system is weakly mixing. This is due to the fact that the matrix coefficients of finite-dimensional (unitary) representations of a group G do not vanish at infinity. Indeed, let π be such a representation. Choosing an orthonormal basis in the space of π, we may realize the operators $\pi(g)$ as hermitian matrices. Since $\pi(g)\pi(g)^*$ is the identity for all $g \in G$, not all coefficients of π vanish at infinity.

We now give a useful characterization of weak mixing.

2.16 Definition. Let G be a group acting in a measure-preserving way on the probability spaces (X_1, μ_1) and (X_2, μ_2). Let $(X_1 \times X_2, \mu_1 \otimes \mu_2)$ be the product of X_1 and X_2 with the product measure $\mu_1 \otimes \mu_2$. The diagonal action of G on $(X_1 \times X_2, \mu_1 \otimes \mu_2)$ is defined by

$$g \cdot (x_1, x_2) = (g \cdot x_1, g \cdot x_2), \quad g \in G, \, x_1 \in X_1, \, x_2 \in X_2.$$

If $(T_i, L^2(X_i, \mu_i))$, $i = 1, 2$ are the associated unitary representations, the representation associated to this action is the tensor product $T_1 \otimes T_2$ on $L^2(X_1 \times X_2, \mu_1 \otimes \mu_2)$.

2.17 Theorem. *The system (X, μ, G) is weakly mixing if and only if the diagonal action of G on $(X \times X, \mu \otimes \mu)$ is ergodic.*

Proof Assume that the system is weakly mixing. Let $k = k(x, y) \in L_0^2(X \times X, \mu \times \mu)$ be G-invariant. We may assume that $k(x, y) = \overline{k(y, x)}$, otherwise consider separately

$$k(x, y) + \overline{k(y, x)} \quad \text{and} \quad i\left(k(x, y) - \overline{k(y, x)}\right).$$

The corresponding integral operator

$$K : L^2(X) \to L^2(X), \quad Kf(x) := \int_X k(x,y)f(y)d\mu(y)$$

is a selfadjoint Hilbert–Schmidt (and, hence, compact) operator, which commutes with all $T_g, g \in G$. Indeed, for any $f \in L^2(X)$,

$$
\begin{aligned}
(K \circ T_g)f(x) &= \int_X k\,(x,y)T_g f(y)d\mu(y) \\
&= \int_X k(x,y)f(g^{-1}y)d\mu\,(y) \\
&= \int_X k(x,g\,y)f(x)d\mu(y) \quad \text{(by invariance of } \mu) \\
&= \int_X k(g^{-1}x,y)f(y)d\mu(y) \quad \text{(by invariance of } k) \\
&= (T_g \circ K)f(x).
\end{aligned}
$$

Let λ be a non-zero eigenvalue of K. The corresponding eigenspace E_λ is finite-dimensional and G-invariant. Hence $E_\lambda = \mathbb{C} \cdot 1$. Thus K is a multiple of the projection onto $\mathbb{C} \cdot 1$. This forces k to be constant almost everywhere.

To show the converse, assume that there exists a finite-dimensional G-invariant subspace V contained in $L_0^2(X)$. Let $\{f_1, f_2, \ldots, f_n\}$ be an orthonormal basis of V and define

$$h : X \times X \to \mathbb{C}, \quad (x,y) \mapsto \sum_{i=1}^n f_i(x)\overline{f_i(y)}.$$

Then $h \in L_0^2(X \times X)$ and

$$
\begin{aligned}
T_g \otimes T_g h(x,y) &= \sum_{i=1}^n T_g f_i(x)\overline{T_g f_i(y)} \\
&= \sum_{i=1}^n \left(\sum_{j=1}^n \langle T_g f_i, f_j \rangle f_j(x) \right) \overline{\left(\sum_{k=1}^n \langle T_g f_i, f_k \rangle f_k(y) \right)}. \\
&= \sum_{i=1}^n \sum_{j=1}^n \sum_{k=1}^n \langle T_g f_i, f_j \rangle \langle f_k, T_g f_i \rangle f_j(x)\overline{f_k(y)}
\end{aligned}
$$

By unitarity of T_g and the G-invariance of V, the set $\{T_g f_1, T_g f_2, \ldots, T_g f_n\}$

is also an orthonormal basis of V. Hence,

$$T_g \otimes T_g h(x,y) = \sum_{j=1}^{n} \sum_{k=1}^{n} \left(\sum_{i=1}^{n} \langle T_g f_i, f_j \rangle \langle f_k, T_g f_i \rangle \right) f_j(x) \overline{f_k(y)}$$

$$= \sum_{j=1}^{n} \sum_{k=1}^{n} \langle f_k, f_j \rangle f_j(x) \overline{f_k(y)}$$

$$= \sum_{j=1}^{n} f_j(x) \overline{f_j(y)}$$

$$= h(x,y).$$

Since $(X \times X, G)$ is ergodic, $h = 0$ almost everywhere and $V = \{0\}$. $\square$

2.18 Example. The action of $G = \mathrm{SL}(n, \mathbb{Z})$ on $\mathbb{T}^n$ is weakly mixing (but not strongly mixing by 2.11 (iv)). Indeed, one has to prove that the action of the diagonal of $\mathrm{SL}(n, \mathbb{Z}) \times \mathrm{SL}(n, \mathbb{Z})$ is ergodic on $\mathbb{T}^n \times \mathbb{T}^n$. This can be shown using the same arguments as for the ergodicity of $\mathrm{SL}(n, \mathbb{Z})$ on $\mathbb{T}^n$ (see Example 1.4 (iii)).

2.19 Remark. We have seen in 2.11 (iv) that the restriction to the upper triangular matrices of the above action of $G = \mathrm{SL}(n, \mathbb{Z})$ on $\mathbb{T}^n$ is not ergodic. Thus, weak mixing is *not* inherited by closed subgroups.

Orbits of Ergodic Actions

We conclude the discussion of ergodicity with some remarks on the structure of the orbits of an ergodic system (X, μ, G), where μ is now allowed to be infinite and *quasi-invariant* (this means that G preserves the measure class of μ). There are two possibilities:
(1) All orbits have measure zero;
(2) There exists one orbit with strictly positive measure (hence its complement has measure zero).

In the first case, one says that the action is *properly ergodic*. If the measure space has no atoms, then every ergodic action of a countable group is properly ergodic (examples are given in 1.4 (ii) and 1.4 (iii)). In the second case, the action is *nearly transitive*.

2.20 Theorem. *Let G be a locally compact group acting continuously on separable metrizable space X. Each of the following statements implies the next one:*
(i) *Every orbit is locally closed (that is, open in its closure);*

(ii) *Endowed with the quotient topology, the orbit space X/G is T_0 (that is, the open sets separate the points of X/G);*

(iii) *There is a sequence $\{E_n\}_n$ of Borel sets which separates the points of X/G;*

(iv) *Every quasi-invariant ergodic Borel measure μ on X is supported by an orbit.*

Proof To show that (i) implies (ii), let $p: X \to X/G$ be the canonical projection. Consider two different orbits Gx and Gy. Then either $\overline{Gx} \supset Gy$ or $\overline{Gx} \cap Gy = \emptyset$. In the second case, the complement of $p(\overline{Gx})$ in X/G is an open set containing $p(Gy)$, but not $p(Gx)$. In the first case, there is an open set $U \subseteq X$ with $U \cap \overline{Gx} = Gx$, since Gx is locally closed. Then, clearly, $G \cdot U \cap \overline{Gx} = Gx$ and $p(G \cdot U)$ is an open set, containing $p(Gx)$, but not $p(Gy)$. Thus, (ii) holds.

To show that (ii) implies (iii), let $\{U_n\}_n$ be a countable base for the topology of X. Then the sets $E_n := p(G \cdot U_n)$ form a countable base for the topology of X/G. Since X/G is T_0, $\{E_n\}_n$ separates the points of X/G.

To show that (iii) implies (iv), let $\{E_n\}_n$ be a separating sequence of Borel sets in X/G. Enlarging it, if necessary, we may assume that the family $\{E_n\}_n$ is closed under taking complements. Let $F_n := p^{-1}(E_n)$. By ergodicity, either F_n or its complement has measure zero. Define

$$H := \bigcap \{F_n \, | \, \mu(F_n) \neq 0\}.$$

Then $\mu(X \setminus H)$ has measure zero. We claim that the G-invariant set H is an orbit. Indeed, suppose that H contains two different orbits Ω_1, Ω_2. Since $\{E_n\}_n$ is separating, for some m, we have $\Omega_1 \in F_m$ and $\Omega_2 \in X \setminus F_m$. As

$$\mu(F_m) \neq 0 \qquad \mu(X \setminus F_m) \neq 0$$

and as $\{F_n\}_n$ is closed under taking complements, this contradicts the definition of H. $\qquad\square$

2.21 Corollary. *Let G be a compact group acting continuously on a separable metrizable space X. Then every invariant ergodic measure μ on X is supported by an orbit.*

2.22 Remark. Actually, all the conditions (i)–(iv) in Theorem 2.20 are equivalent. What remains to prove is that (iv) implies (i). This is a deep result of J. Glimm. For the proof, see [Zi], Theorem 2.1.14.

§3 Invariant Measures and Unique Ergodicity

Let X be a set with a σ-algebra $\mathcal{B}$. Consider a group G acting on X in a measurable way. We shall be concerned here with the question of existence and uniqueness of G-invariant measures on X. So, denote the induced G-action on measures on $(X, \mathcal{B})$ by

$$g \cdot \mu(A) := \mu(g^{-1}(A)), \quad \forall g \in G, A \in \mathcal{B}.$$

Let $\mathcal{M}^1(X, G)$ denote the (possibly empty) set of G-invariant probability measures on $(X, \mathcal{B})$. This is a convex subset of the vector space of all measures on $(X, \mathcal{B})$, and the ergodic measures may be characterized geometrically as follows. Recall that x is called an *extreme point* of a convex set C if it does not lie on a proper line in C, that is, whenever $x = ty + (1-t)z$ for some $y, z \in C$, $t \in [0, 1]$, then $x = y$ or $x = z$.

3.1 Proposition. *The measure* $\mu \in \mathcal{M}^1(X, G)$ *is ergodic if and only if* μ *is an extreme point in* $\mathcal{M}^1(X, G)$.

Proof Let μ be an extreme point and consider a G-invariant measurable set $A \subseteq X$. If $\mu(A) \neq 0$ and $\mu(A) \neq 1$, then the measures defined by

$$\mu'(B) := \frac{\mu(B \cap A)}{\mu(A)}, \quad \mu''(B) := \frac{\mu(B \cap (X \setminus A))}{\mu(X \setminus A)}$$

are in $\mathcal{M}^1(X, G)$ and different from μ. Moreover,

$$\mu = \mu(A)\mu' + \mu(X \setminus A)\mu'',$$

a contradiction.

To show the converse, let μ be ergodic and assume that there are $t \in (0, 1)$ and $\mu', \mu'' \in \mathcal{M}^1(X, G)$ with $\mu = t\mu' + (1-t)\mu''$. Then μ' is absolutely continuous with respect to μ and, by the Radon Nikodym theorem, we find a unique $f \in L^1(\mu)$ with $\mu' = f d\mu$. On the other hand, the invariance of μ' implies that, for all $g \in G$,

$$\mu' = g \cdot \mu' = T_{g^{-1}} f d\mu.$$

Thus, $f = T_{g^{-1}} f$ for all $g \in G$, and f is constant, by Theorem 1.3. Hence, $\mu = \mu'$ and μ is an extreme point. $\Box$

The following theorem shows that the study of the most general G-action may be reduced to the study of ergodic G-actions. The proof may be found in [Va1] for the case of finite measures, and [Da4] shows how this extends to the general case.

3.2 Theorem (Ergodic Decomposition). *Suppose that a locally compact, second countable group G acts continuously on a locally compact and metrizable space X. Let μ be a G-invariant Borel measure on X. Then there exists a measure space (Y, ν), a partition of X into measurable G-invariant subsets $X_y, y \in Y$, and measures μ_y on X_y such that*

(i) *for any measurable set $A \subseteq X$, the set $A \cap X_y$ is μ_y-measurable for almost all $y \in Y$ and $\mu(A) = \int_Y \mu_y (A \cap X_y) \, d\nu(y)$;*

(ii) *for almost all $y \in Y$, the action of G on X_y is ergodic (with respect to the measure μ_y).*

We turn to the question of the existence of invariant measures. Let X be a second countable locally compact space. The space $\mathcal{M}(X)$ of all regular finite measures on X, endowed with the total variation norm, is a Banach space. Recall that $\mathcal{M}(X)$ may be identified with the topological dual space of $C_0(X)$, the Banach space of the continuous functions vanishing at infinity with the supremum norm. This is the standard *Riesz Representation Theorem* which may be stated as follows. The mapping

$$\Psi : \mathcal{M}(X) \to C_0(X)^*, \quad \Psi(\mu)(f) := \int_X f \, d\mu$$

is an isometric, order-preserving isomorphism (for a proof, see e.g. [Rd], 2.14).

When convenient, we shall make no distinction between $\mathcal{M}(X)$ and $C_0(X)^*$. Consider now a group G acting continuously on X. The convex set $\mathcal{M}^1(X, G)$ is a subset of the unit ball in $\mathcal{M}(X) = C_0(X)^*$. Hence, by the Banach–Alaoglu theorem, it is relatively compact with respect to the weak $-*$-topology, and it is compact if X is compact .

3.3 Theorem. *Let X be a second countable compact space and $\varphi : X \to X$ be a continuous map. Then there exists a φ-invariant measure on X.*

Proof Take an arbitrary probability measure μ on X and define

$$\mu_n := \frac{1}{n+1} \sum_{j=0}^{n} \varphi^j \cdot \mu.$$

Then $\{\mu_n\}_n$ is contained in $\mathcal{M}^1(X)$ which is weak $-*$ compact and metrizable (by second countability). Thus we find a convergent subsequence $\{\mu_{n_k}\}_{k \in \mathbb{N}}$ whose limit is clearly φ-invariant. $\qquad\square$

3.4 Remarks.

(i) The same argument as in the previous proof works, of course, for the continuous case $\{\varphi_t\}_{t\geq 0}$. One only has to define the mesure μ_n by

$$\mu_n(f) := \frac{1}{n} \int_0^n \mu(f \circ \varphi_t)dt, \qquad \forall f \in C(X).$$

(ii) A locally compact group G is *amenable*, if there exists a sequence of compact subsets $\{K_n\}_n$ of G such that $\operatorname{int} K_n \subseteq K_{n+1}$ for all $n \in \mathbb{N}$ and

$$\lim_{n\to\infty} \frac{\lambda(gK_n \Delta K_n)}{\lambda(K_n)} = 0$$

uniformly on compact sets, where λ denotes a left invariant Haar measure on G and Δ the symmetric difference.

Suppose G acts on a compact space X. One may consider measures μ_n on X defined by

$$\mu_n(f) := \frac{1}{\lambda(K_n)} \int_{K_n} \mu(T_g f)d\lambda(g), \qquad \forall f \in C(X),$$

and prove as above the existence of G-invariant measures on X. Concerning the theory of amenable groups, see [Pi] and [Pt].

3.5 Corollary. *Let X be a second countable compact space. Let $\varphi : X \to X$ be a continuous mapping. Then there exist φ-ergodic probability measures on X.*

Proof The set $\mathcal{M}^1(X, \varphi)$ is non-empty (by Theorem 3.3), weak $-*$-compact and convex. By the Krein–Milman theorem, there exist extreme points in $\mathcal{M}^1(X, \varphi)$, and the conclusion follows from Proposition 3.1. $\quad\square$

Again, ergodic measures exist for any action of a general amenable group. A situation of particular interest is the following.

3.6 Definition. Let G be a group acting on a Borel space $(X, \mathcal{B})$. The action is called *uniquely ergodic*, if there exists a unique G-invariant probability measure on X (which is automatically ergodic by Proposition 3.1).

In order to characterize unique ergodicity, we need the notion of generic points. Here, we shall only deal with the discrete situation (iteration of a single transformation).

3.7 Definition. Let X be a second countable compact space, $\varphi : X \to X$ be a continuous mapping and $\mu \in \mathcal{M}^1(X, \varphi)$. A point $x \in X$ is called *generic with respect to μ* if, for all $f \in C(X)$,

$$\lim_{n\to\infty} \frac{1}{n+1} \sum_{j=0}^n f(\varphi^j x) = \mu(f). \tag{$*$}$$

Birkhoff's ergodic theorem says that almost all points are generic with respect to an ergodic measure. Indeed, take a countable dense subset D of the unit sphere in $C(X)$. Then, for every $f \in D$, equality $(*)$ holds, except on a set N_f of measure zero. Then $N = \cup_{f \in D} N_f$ has measure zero, and, for all $f \in C(X)$, equality $(*)$ holds outside N.

3.8 Theorem. *Let X be a second countable compact space. Let $\varphi : X \to X$ be a continuous mapping. Then the following are equivalent:*
(i) *For every $f \in C(X)$, there exists a constant $c(f) \in \mathbb{C}$ such that*

$$\lim_{n \to \infty} \frac{1}{n+1} \sum_{j=0}^{n} f(\varphi^j x) = c(f)$$

uniformly in $x \in X$;
(ii) *There exists a measure $\mu \in \mathcal{M}^1(X, \varphi)$ for which every $x \in X$ is generic;*
(iii) *The system (X, φ) is uniquely ergodic.*

Proof Suppose (i) holds. The linear form on $C(X)$ defined by

$$\mu(f) := \lim_{n \to \infty} \frac{1}{n+1} \sum_{j=0}^{n} f(\varphi^j x)$$

is positive and satisfies $|\mu(f)| \leq \|f\|_\infty$, $\mu(1) = 1$. Hence, by the Riesz representation theorem, μ is a regular probability measure which is obviously invariant. Moreover, every $x \in X$ is generic with respect to μ. This shows that (i) implies (ii).

Now let μ be a φ-invariant probability measure for which each point is generic. Let ν be another φ-invariant probability measure on X. The bounded convergence theorem implies that, for all $f \in C(X)$,

$$\int_X f d\nu = \lim_{n \to \infty} \frac{1}{n+1} \int_X \sum_{j=0}^{n} f(\varphi^j x) d\nu(x)$$

$$= \int_X \lim_{n \to \infty} \frac{1}{n+1} \sum_{j=0}^{n} f(\varphi^j x) d\nu(x)$$

$$= \int_X \left(\int_X f d\mu \right) d\nu = \int_X f d\mu.$$

This shows that $\mu = \nu$. Hence, (ii) implies (iii).

Assume that (X, φ) is uniquely ergodic, with unique invariant probability measure μ. Suppose that (i) does not hold. Then there exist $\varepsilon > 0$, $g \in C(X)$ and, for all $n \in \mathbb{N}$, a point $x_n \in X$ and an integer

$N_n \geq n$ such that

$$\left| \frac{1}{N_n + 1} \sum_{j=0}^{N_n} g(\varphi^j x_n) - \int g d\mu \right| \geq \varepsilon.$$

Let ν be any weak $-*$-limit point of the sequence

$$\left\{ \frac{1}{N_n + 1} \sum_{j=0}^{N_n} \delta_{\varphi^j x_j} \right\}_{n \in \mathbb{N}}.$$

As is easily verified, ν is φ-invariant. On the other hand, ν satisfies

$$\left| \int_X g d\nu - \int_X g d\mu \right| \geq \varepsilon.$$

Hence, $\nu \neq \mu$, contradicting the unique ergodicity. This shows that (iii) implies (i). $\qquad \Box$

3.9 Theorem. *Let X be a compact metric space with a probability measure μ whose support is all of X. If $\varphi : X \to X$ is μ-ergodic and isometric, then (X, φ) is uniquely ergodic.*

Proof Let $f \in C(X)$, and define

$$f_n := \frac{1}{n+1} \sum_{j=0}^{n} f \circ \varphi^j.$$

Then, since φ is isometric, $\{f_n\}_n$ is an equicontinuous sequence in $C(X)$. Hence, it has uniformly convergent subsequences, by the Arzela–Ascoli theorem. The limit of such a subsequence is φ-invariant and continuous. Let Y be the subset of X on which this limit coincides with $\int_X f d\mu$. Then Y is closed and, by ergodicity, $\mu(Y) = 1$. Thus, $Y = X$, since X is the support of μ. This shows that $\{f_n\}_n$ converges uniformly and, by Theorem 3.8, that μ is the only φ-invariant measure. $\qquad \Box$

3.10 Example. Consider (see Example 1.4 (ii)) the system on the circle $\mathbb{T}$ given by

$$\varphi(z) = e^{2\pi i \alpha} z, \qquad \forall z \in \mathbb{T}.$$

This system is ergodic for the Lebesgue measure if and only if α is irrational. In this case, the Lebesgue measure is the unique invariant measure. In the case where α is rational, the ergodic measures are the normalized counting measures on the periodic orbits.

Notes

Ergodic theory is, of course, a very broad subject, with a huge literature. The material covered in this chapter is standard and may be found in several places. For an excellent, comprehensive overview, we strongly recommend the book [KH] by A. Katok and B. Hasselblatt. Other standard general references are [CFS], [Hal], [Man], [Wl]. Concerning the relationship between ergodic theory and classical mechanics, see the monograph [AA]. The original proof of Birkhoff's ergodic theorem is in [Bir]. Some deep applications of ergodic theory to combinatorial number theory may be found in [Fu2]. For the point of view of unitary group representations, see [Mc], as well as the first chapters in [Zi].

Chapter II

The Geodesic Flow of Riemannian Locally Symmetric Spaces

One of the most studied flows is the geodesic flow of Riemannian manifolds. It is defined as follows. Let M be a Riemannian manifold with tangent bundle TM, and let

$$T^1 M = \{(p, v) \in TM \mid \|v\| = 1\}$$

be its unit tangent bundle. Each $(p, v) \in T^1 M$ determines a unique geodesic γ through p in the direction v. For $t \in \mathbb{R}$, let p_t be the point on γ at distance t from p in the direction of v and let v_t be the unit vector tangent to γ at the point p_t, in the same direction. Then

$$\varphi_t(p, v) := (p_t, v_t) \qquad t \in \mathbb{R}$$

is a one-parameter group of transformations on $T^1 M$ preserving a natural measure μ.

If the manifold M has positive or zero curvature, the dynamical system is known to be non-ergodic.

For instance, for each point p on the sphere S^2, for a neighbourhood U of p and $\varepsilon > 0$ small enough, the set

$$A = \{\varphi_t(q, v) \mid q \in U, \arg v < \varepsilon, t \in \mathbb{R}\}$$

is an invariant set with $\mu(A) \neq 0$ and $\mu(T^1 S^2 \setminus A) \neq 0$. However, the geometry of manifolds with negative curvature is characterized by a certain instability of the geodesics. For instance, if the curvature at each point lies between two negative bounds, the distance between two infinitesimally near geodesics increases exponentially (at least in one direction). Thus, if the measure of $T^1 M$ is finite, one expects the points on each geodesic to become uniformly distributed. We will discuss this phenomenon later on, giving a precise formulation. In any case, ergodic properties of the geodesic flow reflect some aspects of the geometry of the Riemannian manifold M.

In this chapter, we study the geodesic flow of locally symmetric Riemannian manifolds. It was the discovery of Gelfand and Fomin, later generalized by Mautner, that the geodesic flow on manifolds with constant negative curvature can be studied by means of representation theory of semisimple Lie groups.

We start by discussing the geodesic flow on the unit tangent bundle of a compact surface of constant negative curvature. Such a surface is a closed Riemann surface of genus ≥ 2 and has the hyperbolic plane $\mathbf{H}^2$ as universal covering. So, its fundamental group may be identified with a cocompact lattice in $PSL(2, \mathbb{R})$. Section 1 contains an introduction to hyperbolic geometry. For later use (in Chapter IV, §2), we recall some standard facts about lattices and fundamental domains in Section 2. In Section 3, we study the geodesic flow on compact (or finite volume) Riemannian surfaces. Using a group theoretic description of this flow, we prove its ergodicity. The geodesic flow of general locally symmetric Riemannian manifold is treated in Section 4.

§1 Some Hyperbolic Geometry

We first recall some standard notions from differential geometry needed later, fixing notation at the same time. References for this material are, for instance, [GHL], [Hel] and [War].

Let M be a connected Hausdorff real C^∞-manifold. For $p \in M$, let $T_p M$ be the tangent space p, and denote by TM the tangent bundle of M.

1.1 Definition. A *Riemannian structure* on M is a smooth symmetric 2-form Q on M such that Q_p is an inner product on $T_p M$ for every $p \in M$.

1.2 Examples.

(i) On $M = \mathbb{R}^n$, the standard Riemannian structure is

$$Q := \sum_{i=1}^{n} (dx_i)^2.$$

(ii) Let M be a regular submanifold of $\mathbb{R}^n$, and let $\iota : M \to \mathbb{R}^n$ be the canonical embedding. Then

$$Q|_M := \sum_{i=1}^{n} (dx_i \circ \iota)^2$$

defines a Riemannian structure on M, called the induced Riemannian structure on M.

(iii) Let $S^2 \subseteq \mathbb{R}^3$ be the unit sphere in $\mathbb{R}^3$. A chart on S^2 is given on the complement U of a half great circle through the point $(1,0,0)$ and the north pole by the diffeomorphism

$$F : (0, 2\pi) \times \left(-\frac{\pi}{2}, \frac{\pi}{2} \right) \to U,$$

$$(\varphi, \theta) \longmapsto (\cos(\varphi)\cos(\theta), \sin(\varphi)\cos(\theta), \sin(\theta)).$$

An easy computation shows that the Riemannian structure induced from $\mathbb{R}^3$, is given by $Q|_U = \cos^2(\theta)d\varphi^2 + d\theta^2$.

1.3 Definition. Let (M, Q) be a Riemannian manifold and $p, q \in M$. For a smooth curve $\gamma : [a, b] \to M$ connecting p and q, the length of γ is

$$L(\gamma) := \int_a^b \sqrt{Q_{\gamma(t)}(\dot{\gamma}(t), \dot{\gamma}(t))} \, dt.$$

Now let

$$d(p, q) := \inf L(\gamma),$$

where the infimum is taken over all smooth curves γ connecting p and q. One can show that ϱ is a metric on M defining the original topology on M. Curves which are locally length minimizing are *geodesics*. For instance, the geodesics on the unit sphere S^2 are the great circles.

We now discuss in greater detail the example of the hyperbolic plane which will play an important rôle in these notes. Convenient references for what follows are [Be], [Ka] and [Le]. Consider the *Poincaré upper half plane*

$$\mathbf{H}^2 = \{ z \in \mathbb{C} | \operatorname{Im} z > 0 \},$$

with the Riemannian structure

$$Q := \frac{dx^2 + dy^2}{y^2}.$$

So, the length of a curve

$$\gamma : [0, 1] \to \mathbf{H}^2, \ \gamma(t) = x(t) + i \, y(t)$$

is given by

$$L(\gamma) = \int_0^1 \frac{\sqrt{\left(\frac{dx}{dt}\right)^2 + \left(\frac{dy}{dt}\right)^2}}{y(t)} \, dt,$$

and the norm on the tangent plane $T_z \mathbf{H}^2 \cong \mathbb{C}$ at $z \in \mathbf{H}^2$ is

$$\|\zeta\|_z = \frac{|\zeta|}{\operatorname{Im} z}.$$

Let $\operatorname{PSL}(2, \mathbb{R}) = \operatorname{SL}(2, \mathbb{R})/\{\pm I\}$, the quotient of $\operatorname{SL}(2, \mathbb{R})$ by its centre.

Then $\mathrm{PSL}(2, \mathbb{R})$ acts on $\mathbf{H}^2$ by Möbius (or linear fractional) transformations

$$g.z = \frac{az + b}{cz + d}, \qquad g = \begin{pmatrix} a & b \\ c & d \end{pmatrix} \in \mathrm{SL}(2, \mathbb{R}), \qquad z \in \mathbf{H}^2.$$

That $g \cdot z \in \mathbf{H}^2$, for $g \in \mathrm{SL}(2, \mathbb{R})$ and $z \in \mathbf{H}^2$, follows from

$$\mathrm{Im}(g \cdot z) = \frac{\mathrm{Im}\, z}{|cz + d|^2}.$$

Möbius transformations are transformations of the type

$$z \mapsto \frac{az + b}{cz + d},$$

defined on $\mathbb{C} \cup \{\infty\}$, with the convention that ∞ is mapped to a/c and that $-d/c$ is mapped to ∞. Given two triples

$$(z_1, z_2, z_3) \quad \text{and} \quad (w_1, w_2, w_3)$$

of points in $\mathbb{C} \cup \{\infty\}$, there exists a unique Möbius transformation T with $T(z_i) = w_i$ for $i = 1, 2, 3$ (see [Rd], Chap. 14).

The next lemma is the first crucial fact.

1.4 Lemma. *The above defined action of* $\mathrm{PSL}(2, \mathbb{R})$ *on* $\mathbf{H}^2$ *is isometric.*

Proof The differential $D_z g$ of the diffeomorphism g at the point $z \in \mathbf{H}^2$ is given by

$$D_z g(\zeta) = \frac{1}{(cz + d)^2} \zeta, \qquad \zeta \in T_z \mathbf{H}^2.$$

Hence,

$$\|D_z g(\zeta)\|_{g \cdot z} = \frac{|\zeta|}{\mathrm{Im}\, z} = \|\zeta\|_z,$$

and g preserves the length of any curve γ in $\mathbf{H}^2$. So,

$$d(g \cdot z, g \cdot w) = d(z, w)$$

for all $z, w \in \mathbf{H}^2$. $\qquad \qquad \qquad \square$

1.5 Exercise. Show that the full isometry group of $\mathbf{H}^2$ is generated by $\mathrm{PSL}(2, \mathbb{R})$ and by the involution $z \to -\bar{z}$.

1.6 Lemma. *The geodesics in* $\mathbf{H}^2$ *are the half circles with centre on the* x-*axis or the half-lines which are parallel to the imaginary axis.*

Proof Let $z, w \in \mathbf{H}^2$. Suppose first that $z = ia$ and $w = ib$ with $b > a > 0$. Let $\gamma : [0, 1] \to \mathbf{H}^2$ be a curve from z to w, $\gamma(t) = x(t) + iy(t)$.

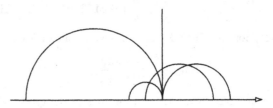

Geodesics in the Poincaré model

Then

$$L(\gamma) = \int_0^1 \frac{\sqrt{\left(\frac{dx}{dt}\right)^2 + \left(\frac{dy}{dt}\right)^2}}{y(t)}\,dt \geq \int_0^1 \frac{\frac{dy}{dt}}{y(t)}\,dt = \log\frac{b}{a}.$$

On the other hand, if one chooses $\gamma(t) = (1-t)ia + tib$, one gets the minimal value $L(\gamma) = \log\frac{b}{a}$. So, the imaginary axis L is a geodesic.

Suppose now that z and w are arbitrary. One can find $g \in \mathrm{SL}(2,\mathbb{R})$ which maps z and w to the imaginary axis. This is done as follows. If z and w have the same real part, take $g \in \mathrm{SL}(2,\mathbb{R})$ such that

$$g \cdot u = u - \mathrm{Re}z, \quad u \in \mathbf{H}^2.$$

Otherwise, consider the circle C through z and w and with centre on the real axis. This circle intersects the real axis in two points t_1, t_2. Now, take $g \in \mathrm{SL}(2,\mathbb{R})$ with

$$g \cdot u = -\frac{1}{u - t_1} + \frac{1}{t_2 - t_1}, \quad u \in \mathbb{C} \cup \{\infty\}.$$

Then g maps t_1 to ∞ and t_2 to 0. Hence, g sends C to the imaginary axis. Since g preserves circles and angles (the euclidean and the hyperbolic angles coincide!), the geodesic $g^{-1}(L)$ through z and w is a circle or a line perpendicular to the real axis.   □

1.7 Lemma. *The hyperbolic distance $d(z,w)$ between two points $z, w \in \mathbf{H}^2$ is given by*

$$\sinh\left(\frac{1}{2}d(z,w)\right) = \frac{|z-w|}{2\sqrt{\mathrm{Im}z\,\mathrm{Im}w}}.$$

Proof As $d(ia, ib) = \log\frac{b}{a}$ for $b > a > 0$ (see the proof of the previous lemma), the formula is readily verified if z and w are on the imaginary axis.

Let z, w be arbitrary points in $\mathbf{H}^2$. Let $g \in \mathrm{SL}(2, \mathbb{R})$ be such that gz and gw are on the imaginary axis. Since g is an isometry,

$$\sinh\left(\frac{1}{2}d(z,w)\right) = \sinh\left(\frac{1}{2}d(gz,gw)\right).$$

On the other hand, as the formula is true for points on the imaginary axis,

$$\sinh\left(\frac{1}{2}d(gz,gw)\right) = \frac{|gz - gw|}{2\sqrt{\mathrm{Im}(gz)\mathrm{Im}(gw)}}.$$

Writing $g = \begin{pmatrix} a & b \\ c & d \end{pmatrix}$ with $ad - bc = 1$, one has

$$|gz - gw| = \frac{|z - w|}{|cz + d||cw + d|} = \frac{|z - w|\sqrt{\mathrm{Im}(gz)\mathrm{Im}(gw)}}{\sqrt{\mathrm{Im}z\mathrm{Im}w}},$$

and this proves the formula. $\qquad\qquad\square$

We shall often make no (notational) distinction between a matrix from $\mathrm{SL}(2,\mathbb{R})$ and the corresponding Möbius transformation or its equivalence class in $\mathrm{PSL}(2,\mathbb{R})$.

1.8 Exercise. Show that

$$d(z,w) = \log\left(\frac{|z - \bar{w}| + |z - w|}{|z - \bar{w}| - |z - w|}\right)$$

for the hyperbolic distance d on $\mathbf{H}^2$.

1.9 Remark. Other models for the hyperbolic plane are obtained by the Riemann mapping theorem. Let $B \subsetneq \mathbb{C}$ be a non-empty simply connected domain and $f : B \to \mathbf{H}^2$ be a biholomorphic mapping of B onto $\mathbf{H}^2$. Then the 2-form Q^B defined by

$$Q^B_p(X_p, Y_p) := Q_{f(p)}(df_p(X_p), df_p(Y_p)), \qquad p \in B,$$

where X, Y are smooth vector fields on B and df_p denotes the differential of f at the point p, turns B into a Riemannian manifold, isometric to $(\mathbf{H}^2, Q)$. If $\mathbb{D}$ denotes the unit disc and

$$f : \mathbb{D} \to \mathbf{H}^2, \quad z \mapsto -i\frac{z+1}{z-1},$$

one finds

$$Q^{\mathbb{D}} = 2\frac{dx^2 + dy^2}{\left(1 - x^2 - y^2\right)^2}.$$

This is the *Poincaré disc model* of the hyperbolic plane.

§2 Lattices and Fundamental Domains

We recall some generalities concerning lattices in locally compact groups and discrete subgroups of $\mathrm{SL}(2,\mathbb{R})$.

Let G be a locally compact group with left invariant Haar measure ν_G. Let H be a closed subgroup of G with Haar measure ν_H. Let Δ_G and Δ_H be the corresponding modular functions on G and H. Recall that Δ_G, a continuous homomorphism from G to $\mathbb{R}^+$, is such that

$$d\nu_G(x^{-1}) = \Delta(x)d\nu_G(x).$$

If $\Delta_G = 1$, then G is said to be *unimodular*. In this case, ν_G is also right invariant.

Assume that $\Delta_G|_H = \Delta_H$. Then there is, up to a scalar multiple, a unique G–invariant Borel measure ν_X on the homogeneous space $X = G/H$. After suitable normalization of the Haar measures, the so-called *Weil formula*

$$\int_G f(x)d\nu_G(x) = \int_X \int_H f(xh)d\nu_H(h)d\nu_X(\dot{x}) \qquad (\dot{x} = xH)$$

holds for all integrable functions f on G (see [We], Chap. II or [Re], Chap. 3). This may be extended to an arbitrary closed subgroup H of G as follows. There exists a quasi-invariant Borel measure ν_X on the homogeneous space $X = G/H$ such that, after suitable normalization of the Haar measures, Weil's formula above holds for all integrable functions f on G (see [Re], Chap. 8).

2.1 Definition. A *lattice* in G is a discrete subgroup Γ in G such that G/Γ carries a finite G–invariant Borel measure.

Observe that a group G containing a lattice Γ is necessarily unimodular. Indeed, $\Delta_\Gamma = 1$, as Γ is discrete. So, Γ is contained in the normal subgroup $\ker \Delta_G$ of G. Hence, the Haar measure of the locally compact group $G/\ker \Delta_G$ is finite. As is easy to prove, this implies that $G/\ker \Delta_G$ is compact (see [We], Chap. II, §8.). Hence, $G/\ker \Delta_G$ is topologically isomorphic to a compact subgroup of $\mathbb{R}^+$. The only such subgroup being trivial, it follows that $G = \ker \Delta_G$.

A first and trivial example is the lattice $\mathbb{Z}^n$ in $\mathbb{R}^n$. More generally, if Γ is a discrete and cocompact subgroup of a locally compact group G, then Γ is a lattice in G. Such lattices are called *uniform* lattices.

There are examples of non-uniform lattices. For instance, $SL(2,\mathbb{Z})$ is a lattice in $SL(2,\mathbb{R})$. This classical example will be treated below in some details.

We first give a characterization of the discrete subgroups of $SL(2,\mathbb{R})$.

2.2 Definition.

(i) Let Γ be a group acting on a locally compact space X. Then Γ is said to act *properly discontinuously*, if for any compact subsets K and L of X the set

$$\{g \in \Gamma |\ gK \cap L \neq \emptyset\}$$

is finite.

(ii) Classically, a discrete subgroup of $PSL(2,\mathbb{R})$ is called a *Fuchsian group*.

2.3 Lemma. *An infinite subgroup Γ of $PSL(2,\mathbb{R})$ is a Fuchsian group if and only if Γ acts properly discontinuously on $\mathbf{H}^2$.*

Proof Suppose that Γ is not discrete. Then there exists a sequence $\{\gamma_n\}_n$ in Γ with $\gamma_n \to e$ and $\gamma_n \neq e$. Then, for any $z \in \mathbf{H}^2$, $\gamma_n z \to z$. Hence, Γ cannot act properly discontinuously on $\mathbf{H}^2$.

As to the converse, it suffices to prove the following. Let $z_0 \in \mathbf{H}^2$, and let K be a compact subset of $\mathbf{H}^2$. Then the set

$$E = \{g \in SL(2,\mathbb{R}) |\ gz_0 \in K\}$$

is compact.

Clearly, E is closed. Viewing $SL(2,\mathbb{R})$ as a subset of $\mathbb{R}^4$, we have to show that E is bounded. As K is compact, there exist constants $M, \delta > 0$ such that

$$|z| \leq M, \qquad \mathrm{Im}\, z \geq \delta \qquad \forall z \in K.$$

Let

$$g = \begin{pmatrix} a & b \\ c & d \end{pmatrix} \in E.$$

Then

$$\left| \frac{az_0 + b}{cz_0 + d} \right| \leq M, \quad \mathrm{Im}\left(\frac{az_0 + b}{cz_0 + d} \right) = \frac{\mathrm{Im}\, z_0}{|cz_0 + d|^2} \geq \delta.$$

Thus, setting $\alpha = \mathrm{Im}\, z_0$,

$$|cz_0 + d|^2 \leq \alpha/\delta, \qquad |az_0 + b|^2 \leq M\alpha/\delta.$$

The set of all $(a, b, c, d) \in \mathbb{R}^4$ satisfying these inequalities is obviously bounded. $\square$

Fundamental Domains for Fuchsian Groups

We shall need some facts about fundamental domains for lattices in $SL(2, \mathbb{R})$ in Chap. IV, §2.

2.4 Definition. Let X be a locally compact space on which a group G acts by homeomorphisms. An open subset F of X is a *fundamental domain* for the action of G if

(i) $\bigcup_{\gamma \in G} \gamma \overline{F} = X$, where $\overline{F}$ is the closure of F, and

(ii) $\gamma_1 F \cap \gamma_2 F = \emptyset$, for all $\gamma_1, \gamma_2 \in G$, $\gamma_1 \neq \gamma_2$.

For instance, $F = \{(x_1, \ldots, x_n) \in \mathbb{R}^n | \, 0 < x_i < 1, \quad i = 1, \ldots, n\}$ is a fundamental domain for the action of $\mathbb{Z}^n$ by translation on $\mathbb{R}^n$.

Let Γ be an infinite Fuchsian group. A fundamental domain for the action of Γ on $\mathbf{H}^2$ is constructed as follows. Choose a point $p \in \mathbf{H}^2$ which is not fixed by any $\gamma \in \Gamma \setminus \{e\}$. Such a point exists. Indeed, Γ is countable and every $g \in SL(2, \mathbb{R})$, $g \neq e$, has at most 2 fixed points in $\mathbf{H}^2$.

2.5 Definition (Dirichlet region). The set

$$D = \{z \in \mathbf{H}^2 | \, d(z, p) < d(z, \gamma p), \quad \forall \gamma \in \Gamma \setminus \{e\}\},$$

where d denotes the hyperbolic metric on $\mathbf{H}^2$ associated to

$$ds^2 = \frac{dx^2 + dy^2}{y^2},$$

is called a *Dirichlet region* for Γ.

Such a region, being the intersection of the hyperbolic half planes

$$H(\gamma) = \{z \in \mathbf{H}^2 | \, d(z, p) < d(z, \gamma p)\},$$

is a connected hyperbolically convex region (that is, it contains with any pair of points z, w the geodesic segment $[z, w]$ connecting them). Its boundary consists of the geodesic arcs

$$L(\gamma) = \{z \in \mathbf{H}^2 | \, d(z, p) = d(z, \gamma p)\},$$

for $\gamma \in \Gamma \setminus \{e\}$. The crucial property of such a domain is the following.

2.6 Lemma. *The Dirichlet region D is a fundamental domain for the action of the Fuchsian group Γ on $\mathbf{H}^2$.*

Proof We first claim that D is open. Indeed, otherwise there exist $z \in D$ and sequences $\{z_n\}_n \in \mathbf{H}^2, \{\gamma_n\}_n \in \gamma \in \Gamma \setminus \{e\}$ such that

$$z_n \to z \quad \text{and} \quad d(z_n, p) \geq d(z_n, \gamma_n p).$$

The sequence $\{\gamma_n p\}$ is then bounded. Since Γ acts properly discontinuously, the sequence $\{\gamma_n\}_n$ is finite. Hence, upon passing to a subsequence if necessary,

$$d(z, p) = \lim d(z_n, p) \geq \lim d(z_n, \gamma_n p) = d(z, \gamma p),$$

for some $\gamma \neq e$. This contradicts the fact that $z \in D$.

Fix $z_0 \in \mathbf{H}^2$. The Γ-orbit of z_0 being discrete, there exists $z \in \Gamma z_0$ such that

$$d(z, p) \leq d(\gamma z_0, p) = d(z_0, \gamma^{-1} p)$$

for all $\gamma \in \Gamma$. So, z is in the closure of D. Indeed, the segment $[p, z)$ is contained in D, as $p \in D$ and $d(z, p) \leq d(z, \gamma p)$ for all $\gamma \in \Gamma$. This shows that $z \in \bar{D}$. Hence, the closure of D meets every Γ-orbit.

Let $z_1 = \gamma z_0$, for $\gamma \in \Gamma, \gamma \neq e$. Assume that both z_0 and z_1 lie in D. Then

$$d(z_0, p) < d(z_0, \gamma^{-1} p) = d(z_1, p),$$

and, similarly, $d(z_1, p) < d(z_0, p)$. This is a contradiction. $\qquad\square$

2.7 Example. Let $\Gamma = \mathrm{PSL}(2, \mathbb{Z})$, the so-called *modular group*. It is easily seen that Γ is a discrete subgroup of $\mathrm{PSL}(2, \mathbb{R})$. Let $p = ni$ for some $n > 1$. Then $\gamma p \neq p$ for all $\gamma \in \Gamma \setminus \{e\}$. Indeed, if

$$\frac{ani + b}{cni + d} = ni \quad \text{for} \quad a, b, c, d \in \mathbb{Z}, \quad ad - bc = 1,$$

then

$$\frac{n}{cn^2 + d^2} = \mathrm{Im}\left(\frac{ani + b}{cni + d}\right) = n,$$

and hence $c = 0$ and $d = 1$ or -1. So, $a = 1$ or -1 and $b = 0$. Whence, $\gamma p = p$ only for $\gamma = e$.

Let P be the domain

$$P = \{z \in \mathbf{H}^2 | \ |z| > 1, |\mathrm{Re} z| < 1/2\}.$$

We claim that P is a fundamental domain for Γ. This will follow from Lemma 2.6 once we have shown that P is the Dirichlet domain D corresponding to Γ and p.

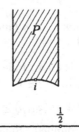

A fundamental domain for the modular group

Let

$$\gamma_1 = \begin{pmatrix} 1 & 1 \\ 0 & 1 \end{pmatrix}, \quad \gamma_2 = \begin{pmatrix} 0 & 1 \\ -1 & 0 \end{pmatrix}.$$

So, $\gamma_1 z = z + 1$, and $\gamma_2 z = -1/z$.

The sides of P are the three geodesic segments $L(\gamma_1), L(\gamma_1^{-1}), L(\gamma_2)$. This shows that D is contained in P.

Assume, by contradiction, that D is strictly contained in P. Then there exist $z \in P$ and

$$\gamma = \begin{pmatrix} a & b \\ c & d \end{pmatrix} \in \mathrm{SL}(2, \mathbb{Z}), \quad \gamma \neq \pm I,$$

such that $\gamma z = \frac{az+b}{cz+d}$ lies in P.

Then $c \neq 0$. Indeed, otherwise $a, d = \pm 1, b \neq 0$ and

$$|\mathrm{Re}\gamma z| = |\mathrm{Re}z + b| \geq |b| - |\mathrm{Re}z| > |b| - \frac{1}{2} \geq \frac{1}{2},$$

so that $\gamma z \notin P$.

As $|z| > 1$ and $|\mathrm{Re}z| < 1/2$, we have

$$\begin{aligned} |cz + d|^2 &= c^2|z|^2 + 2cd\,\mathrm{Re}z + d^2 \\ &> c^2 + d^2 - |cd| \\ &= (|c| - |d|)^2 + |cd|. \end{aligned}$$

Now $(|c| - |d|)^2 + |cd| \neq 0$, since otherwise $c = d = 0$, which is impossible. As $(|c| - |d|)^2 + |cd|$ is an integer, we have $|cz + d|^2 > 1$. Hence,

$$\mathrm{Im}\gamma z = \frac{\mathrm{Im}z}{|cz + d|^2} < \mathrm{Im}z.$$

Replacing z by γz and γ by γ^{-1} in the above argument shows that one also has

$$\mathrm{Im}z < \mathrm{Im}\gamma z.$$

This is a contradiction.

2.8 Example. The image Q of the domain P above by the transformation $z \mapsto -1/z$ is again a fundamental domain for the modular group $PSL(2, \mathbb{Z})$.

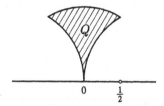

Another fundamental domain for the modular group

2.9 Exercise. Let $\Gamma(2)$ be the subgroup of $PSL(2, \mathbb{Z})$ consisting of all matrices which are congruent to the identity I modulo 2. This is a so-called congruence subgroup. It can be shown that it is generated by the matrices $\begin{pmatrix} 1 & 2 \\ 0 & 1 \end{pmatrix}$ and $\begin{pmatrix} 1 & 0 \\ 2 & 1 \end{pmatrix}$. See, for instance, [Le], Chap. VII, 6C. Show that the domain

$$\{z \in \mathbf{H}^2 | \; |z| < 1, |z - \tfrac{1}{2}| > \tfrac{1}{2}, |z + \tfrac{1}{2}| > \tfrac{1}{2}\}$$

is a fundamental domain for $\Gamma(2)$.

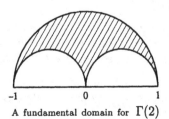

A fundamental domain for $\Gamma(2)$

We turn to the question of the existence of finite invariant measures.

2.10 Lemma. *The measure* $y^{-2}dxdy$ *on* $\mathbf{H}^2$ *is invariant under the action of* $SL(2,\mathbb{R})$.

Proof For $g = \begin{pmatrix} a & b \\ c & d \end{pmatrix} \in SL(2,\mathbb{R})$, denote by M_g the associated Möbius transformation. The Jacobian of M_g at $z \in \mathbf{H}^2$, viewed as a linear transformation on $\mathbb{R}^2$, is equal to $|cz + d|^{-4}$. Thus, by change of variables, we find

$$\int_{\mathbf{H}^2} f\left(g^{-1} \cdot z\right) \frac{dxdy}{y^2} = \int_{\mathbf{H}^2} f(z) \frac{1}{|cz+d|^4} \frac{|cz+d|^4}{y^2} dx\, dy$$

$$= \int_{\mathbf{H}^2} f(z) \frac{dx\, dy}{y^2}.$$

$\square$

Let Γ be a group acting on a measure space (X, μ). We say that a subset F of X is a *Borel fundamental domain* for the action of Γ if F is measurable and satisfies the following two conditions:

(i) $\mu\left(X \setminus \bigcup_{\gamma \in \Gamma} \gamma F\right) = 0$ and

(ii) $\mu(\gamma_1 F \cap \gamma_2 F) = 0$, for all $\gamma_1, \gamma_2 \in \Gamma$, $\gamma_1 \neq \gamma_2$.

2.11 Lemma. *Suppose that a unimodular locally compact group* G *acts transitively on a locally compact space* X *and that the stabilizer* K *of a point* $x_0 \in X$ *is compact. Let* μ *be a* G*-invariant Borel measure on* $X \cong G/K$. *Let* Γ *be a discrete subgroup of* G. *Assume that there exists a Borel fundamental domain* F *for the action of* Γ *on* X *with finite measure. Then* Γ *is a lattice in* G.

Proof We may assume that $X = G/K$ and that $x_0 = K \in F$. Consider the linear form ν on $C_c(\Gamma \backslash G)$, the space of all continuous functions with compact support on $\Gamma \backslash G$, defined by

$$\nu(f) = \int_F d\mu(\dot{x}) \int_K f(\pi(x)k)dk, \qquad \forall f \in C_c(\Gamma \backslash G), \ (\dot{x} = xK)$$

where dk is a Haar measure on K and π is the canonical mapping from G onto $\Gamma \backslash G$. As $\mu(F) < \infty$, the Radon measure ν on $\Gamma \backslash G$ is finite. Moreover, ν is G-invariant. Indeed, for $f \in C_c(G)$, define $Tf \in C_c(\Gamma \backslash G)$ by

$$Tf(\pi(x)) = \sum_{\gamma \in \Gamma} f(\gamma x), \qquad \forall x \in G.$$

Then, by the invariance of μ and Weil's formula,

$$\nu(Tf) = \sum_{\gamma \in \Gamma} \int_F d\mu(\dot{x}) \int_K f(\gamma x k)dk$$

$$= \sum_{\gamma \in \Gamma} \int_{\gamma^{-1}F} d\mu(\dot{x}) \int_K f(x k)dk$$

$$= \int_{G/K} d\mu(\dot{x}) \int_K f(x k)dk = \int_G f(g)dg,$$

where dg is a (left) Haar measure on G. As T commutes with right translations $\lambda(y)$ by elements y in G and as dg is also right invariant, this implies that $\nu(\lambda(y)Tf) = \nu(Tf)$. Since

$$C_c(\Gamma\backslash G) = \{Tf \mid f \in C_c(G)\}$$

(see [We], Chap. II, §9 or [Re], Chap. 3, 4.2), this proves the claim. □

2.12 Example. We show that $\mathrm{SL}(2,\mathbb{Z})$ is a lattice in $\mathrm{SL}(2,\mathbb{R})$.

Let $\mathrm{SL}(2,\mathbb{R})$ act by Möbius transformations on the upper half plane $\mathbf{H}^2$ with the invariant measure $dxdy/y^2$. This is a transitive action, and $\mathrm{SO}(2)$ is the stabilizer of the point i. So, $\mathbf{H}^2$ may be identified with $\mathrm{SL}(2,\mathbb{R})/\mathrm{SO}(2)$. As $\mathrm{SO}(2)$ is compact, it suffices to show that there exists a measure theoretical fundamental domain with finite measure for the action of $\mathrm{SL}(2,\mathbb{Z})$ on $\mathbf{H}^2$.

As shown above, the set

$$F = \{z \in \mathbf{H}^2 \mid |z| > 1 \quad \text{and} \quad \mathrm{Re}(z) < 1/2\}$$

is a fundamental domain for this action. The area of F is finite. In fact, by the Gauß–Bonnet formula (see Propositon 2.13 below), the area of F is equal to $\pi/3$. This may also be seen by direct computation:

$$\mathrm{area}(F) = \int_F \frac{dxdy}{y^2} = \int_{-1/2}^{1/2} \int_{\sqrt{1-x^2}}^{\infty} \frac{dxdy}{y^2} = \int_{-1/2}^{1/2} \frac{1}{\sqrt{1-x^2}}dx = \frac{\pi}{3}.$$

Let $\overline{\mathbf{H}^2} = \mathbf{H}^2 \cup \partial\mathbf{H}^2$, where $\partial\mathbf{H}^2 = \mathbb{R} \cup \{\infty\}$ denotes the *boundary* of $\mathbf{H}^2$.

Let D be a Dirichlet region for a Fuchsian group Γ. Recall that the boundary of D consists of geodesic segments. The intersection of two such segments is a *vertex* of D. Observe that some vertices of D may lie on $\mathbb{R} \cup \{\infty\}$. Such vertices are called *vertices at infinity*. For instance, ∞ is a vertex at infinity for the fundamental region of $\mathrm{PSL}(2,\mathbb{Z})$ as in Example 2.7. For the region in Example 2.8, 0 is a vertex at infinity. The vertices at infinity of the fundamental region for the congruence subgroup $\Gamma(2)$ as in Exercise 2.9 are $-1, 0$ and 1.

2.13 Proposition (Gauß–Bonnet Formula). *The area of a hyperbolic triangle Δ in $\mathbf{H}^2$ with vertices in $\overline{\mathbf{H}^2}$ is*

$$\pi - (\alpha + \beta + \delta),$$

where α, β, δ are the angles at the vertices of Δ. (Notice that the angle at a vertex at infinity is zero.) More generally, let P be a (hyperbolic) polygon with n sides in $\mathbf{H}^2$ and vertices in $\overline{\mathbf{H}^2}$. Let $\alpha_1, \cdots, \alpha_n$ denote the angles at the vertices of P. Then the area of $\mathbf{H}^2$ is

$$(n - 2)\pi - (\alpha_1 + \ldots + \alpha_n).$$

Proof Since a polygon with n sides, $n \geq 3$, may be decomposed into $n - 2$ adjacent triangles, it suffices to prove the formula for a triangle Δ.

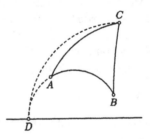

Suppose first that Δ has a vertex at infinity. Applying a suitable transformation from $\mathrm{SL}(2, \mathbb{R})$, we may assume that one vertex A is ∞ and that the other two vertices B, C lie on the unit circle. So, the angle at A is 0. Let α, β denote the angles at B, C. Then the area of Δ is

$$\int_{\cos(\pi - \alpha)}^{\cos \beta} \left(\int_{\sqrt{1-x^2}}^{\infty} \frac{dy}{y^2} \right) dx = \pi - (\alpha + \beta).$$

Suppose now that the vertices A, B, C of Δ are in $\mathbf{H}^2$. Let D be a point of the intersection of the geodesic connecting A and B with the real axis. Then the area of Δ is the difference of the areas of the triangles ACD and CBD. Hence, it is equal to

$$(\pi - (\alpha + \gamma + \theta)) - (\pi - \theta - (\pi - \beta)) = \pi - (\alpha + \beta + \gamma),$$

where α, β, γ are the angles of Δ at A, B, C and θ is the angle of CBD at C. $\qquad\square$

Structure of Dirichlet Regions

We record some classical facts about the structure of a Dirichlet region for a lattice Γ acting on the upper half plane $\mathbf{H}^2$.
Let D be a Dirichlet region for Γ. Then D has finite hyperbolic area.

2.14 Theorem (Siegel). *A lattice Γ is geometrically finite, that is, any Dirichlet region for Γ has finitely many sides.*

Proof Let D be a Dirichlet region for Γ.

First step: D has only finitely many vertices at infinity. Indeed, given n vertices at infinity, thanks to the convexity of D, the polygon with these vertices lies in D. This polygon has area $\pi(n-2)$, by the Gauß–Bonnet formula (Proposition 2.13). Since the area of D is finite, this proves the claim.

Second step: The boundary ∂D of D in $\mathbf{H}^2$ has only finitely many connected components. Indeed, any such component is a maximal connected union of geodesic segments between vertices of D lying in $\mathbf{H}^2$. Let C be such a component. Two cases may occur: either C is a closed polygon in $\mathbf{H}^2$, in which case $C = \partial D$, or the vertices on C tend (from each side) to a vertex at infinity. But there are only finitely many vertices at infinity and to a vertex at infinity correspond two connected components. This proves the claim.

Third step: Fix a connected component C of ∂D in $\mathbf{H}^2$. Let x_k be the vertices on C and let $0 < \omega_k < \pi$ be the angle at x_k. We claim that $\omega_k \leq 3\pi/4$ for only finitely many vertices x_k.

Indeed, let p be the base point in $\mathbf{H}^2$ defining D. Let Δ_k be the triangle with vertices x_k, x_{k+1}, p and angles $\alpha_k, \beta_k, \gamma_k$, respectively. So, $\omega_k = \alpha_k + \beta_{k-1}$. By convexity, Δ_k lies in D. We have, for any n,

$$\text{area}(\Delta_k) = \pi - (\alpha_k + \beta_k + \gamma_k)$$

and so

$$\sum_{k=0}^{n} \text{area}(\Delta_k) = \pi - \alpha_0 - \beta_n + \sum_{k=1}^{n}(\pi - \omega_k) - \sum_{k=0}^{n}\gamma_k.$$

Since $\sum_{k=0}^{n}\text{area}(\Delta_k) \leq \text{area}(D) < \infty$, and since $\sum_{k=0}^{n}\gamma_k \leq 2\pi$, this implies that

$$\sum_{k=1}^{n}(\pi - \omega_k) + \pi - \beta_n \leq \alpha_0 + 2\pi + \text{area}(D) < \infty.$$

As $\beta_n, \omega_k < \pi$, this shows that the series $\sum_{k=1}^{\infty}(\pi - \omega_k)$ converges. In particular, $\lim \omega_k = \pi$ as $k \to \infty$ and this proves the claim.

Fourth step: D has only finitely many vertices. To see this, decompose the set of vertices into a union of congruent vertices under Γ. For a vertex x_k of D, let $\gamma_1 x_k = x_k, \gamma_2 x_k, \ldots$ be the distinct vertices congruent to x_k. Let $\omega_1, \omega_2, \ldots$ be the angles at $\gamma_1 x_k, \gamma_2 x_k, \ldots$ ($0 < \omega_i < \pi$).

Now x_k is a vertex of $\gamma_i^{-1} D$ with angle ω_i and $\gamma_i^{-1} D \cap \gamma_j^{-1} D = \emptyset$. Hence,

$$\omega_1 + \omega_2 + \cdots \leq 2\pi.$$

As, by the above, only finitely many of the angles ω_k are $\leq 3\pi/4$, this implies that there are only finitely many vertices $\gamma_1 x_k, \ldots, \gamma_n x_k$ congruent to x_k.

Let Γ_k be the stabilizer of x_k in Γ, and let m be the order of Γ_k. (Observe that Γ_k is finite as it is a discrete subgroup of a conjugate of $SO(2)$.) The translates of D which have x_k as vertex are all of the form

$$\gamma \gamma_i^{-1} D, \qquad \gamma \in \Gamma_k, \quad 1 \leq i \leq n.$$

Since these translates are pairwise disjoint, this implies that

$$2\pi = m\omega_1 + \cdots + m\omega_n.$$

Clearly, such a partition of 2π is impossible with all angles $\omega_i > 3\pi/4$. Hence, $\omega_i \leq 3\pi/4$ for at least one i. But (by the third step) there are only finitely many vertices with angle $\leq 3\pi/4$. Hence, there are only finitely many orbits of vertices of D. As the number of vertices congruent to a vertex is finite (see above), the claim is proved. $\square$

Clearly, if the whole boundary of D is contained in $\mathbf{H}^2$, then D is relatively compact in $\mathbf{H}^2$ and, hence, Γ is cocompact in $PSL(2, \mathbb{R})$. In fact, a stronger result is true:

2.15 Proposition. *If Γ is a non-cocompact lattice, then any Dirichlet domain D has at least one vertex at infinity.*

Proof In order to see this, let p be the point used to construct D. Look at geodesic rays starting from p. If the intersection of any such geodesic with D has a bounded hyperbolic length, then D is relatively compact in $\mathbf{H}^2$. So, some geodesic ray has infinite length. The intersection s of this ray with the boundary of D is a point at infinity (that is in $\mathbb{R} \cup \{\infty\}$). Since Γ has finite area, the intersection of the boundary of D and $\mathbb{R} \cup \{\infty\}$ contains no interval. This clearly implies that s is a vertex at infinity. $\square$

The action of $PSL(2, \mathbb{R})$ on $\mathbf{H}^2$ extends to an action on its boundary $\mathbb{R} \cup \{\infty\}$ by Möbius transformations. Möbius transformations are classified as follows.

2.16 Definition.

(i) An element $g \in \mathrm{PSL}(2,\mathbb{R})$ is *parabolic* if g has exactly one fixed point in $\mathbb{R} \cup \{\infty\}$. Thus, g is parabolic if and only if g is conjugate to an element from N, the subgroup of $\mathrm{PSL}(2,\mathbb{R})$ consisting of the strictly upper triangular matrices. In other words, g is parabolic if the corresponding transformation on $\mathbf{H}^2$ is conjugate to a translation parallel to the real axis.

(ii) g is *elliptic* if g has a fixed point in $\mathbf{H}^2$, that is, if g is conjugate to an element in $\mathrm{SO}(2)/\{\pm I\}$.

(iii) g is *hyperbolic* if g has two fixed points on $\mathbb{R} \cup \{\infty\}$, that is, if g is conjugate to a diagonal matrix.

Any $g \in \mathrm{PSL}(2,\mathbb{R})$ is either elliptic, parabolic or hyperbolic.

2.17 Proposition. *Let Γ be a non-cocompact lattice, and let D be a Dirichlet domain for Γ. Let x be a vertex at infinity of D. Then x is fixed by a parabolic element γ in Γ which may be chosen so that the stabilizer of x in Γ is $\langle\gamma\rangle$, the cyclic subgroup generated by γ.*

Proof Let $g \in \mathrm{PSL}(2,\mathbb{R})$ be such that $x = g\infty$. Upon replacing Γ by $g^{-1}\Gamma g$ and D by $g^{-1}D$, we may assume that $x = \infty$.

There are infinitely many $\gamma \in \Gamma$ such that ∞ is a vertex of γD. Indeed, D has two sides of the form $\{z \in \mathbf{H}^2 | \operatorname{Re} z = \alpha\}$ and the translates of $\overline{D}$ by elements from Γ cover the whole of $\mathbf{H}^2$.

Let $\gamma \in \Gamma$ be such that ∞ is a vertex of γD. Then, obviously, $\gamma^{-1}\infty$ is a vertex at ∞ of D. But, by Siegel's theorem (Theorem 2.14), D has only finitely many vertices. Hence, there are two (in fact, infinitely many) elements $\gamma_1, \gamma_2 \in \Gamma$, $\gamma_1 \neq \gamma_2$ such that

$$\gamma_1^{-1}\infty = \gamma_2^{-1}\infty.$$

That is, $\gamma_2\gamma_1^{-1}$ fixes ∞.

Let $\gamma \in \Gamma$ with $\gamma\infty = \infty$. We claim that γ is parabolic. First, γ is of the form

$$\gamma z = az + b, \qquad \forall z \in \mathbf{H}^2$$

for some $a \in \mathbb{R}^+$, $b \in \mathbb{R}$. Assume by contradiction, that γ is not parabolic, that is, $a \neq 1$. Let $p \in D$ be the point used in the construction of D. Then, since $\operatorname{Im}\gamma p = a\operatorname{Im} p$,

$$\operatorname{Im}\gamma p \neq \operatorname{Im} p.$$

Replacing γ by γ^{-1} if necessary, we may assume that $\operatorname{Im}\gamma p > \operatorname{Im} p$.

Let $q \in \mathbf{H}^2$ be the point with

$$\operatorname{Im} q = \operatorname{Im} p \quad \text{and} \quad \operatorname{Re} q = \operatorname{Re} \gamma p,$$

and consider the half lines (geodesics)

$$\{p + it \mid t \geq 0\} \quad \text{and} \quad \{q + it \mid t \geq 0\}.$$

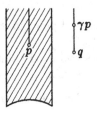

One has (see Lemma 1.7)

$$\sinh \frac{1}{2} d(p + it, q + it) = \frac{|p - q|}{2(t + \operatorname{Im} p)},$$

so that $\lim_{t \to \infty} d(p + it, q + it) = 0$.

On the other hand, for $t > \operatorname{Im} \gamma p - \operatorname{Im} p$, one has $d(q, q + it) = d(q, \gamma p) + d(q + it, \gamma p)$, and, since $d(q, q + it) = d(p, p + it)$,

$$\begin{aligned}
d(p + it, \gamma p) &\leq d(p + it, q + it) + d(q + it, \gamma p) \\
&= d(p + it, q + it) + d(q, q + it) - d(q, \gamma p) \\
&= d(p + it, q + it) + d(p, p + it) - d(q, \gamma p).
\end{aligned}$$

So, as $q \neq \gamma p$, $d(p + it, \gamma p) < d(p + it, p)$ for t sufficiently large. But $p + it$ lies in D (for t large enough). This means that

$$d(p + it, \delta p) > d(p + it, p), \qquad \forall \delta \in \Gamma, \ \delta \neq e.$$

This is a contradiction.

Hence, the stabilizer of ∞ in Γ is a subgroup of the group $N \cong \mathbb{R}$ of the translations $z \mapsto z + x$, $x \in \mathbb{R}$. It is cyclic since it is discrete. $\square$

We are going to show that the converse to Proposition 2.17 above is also true.

2.18 Lemma. *Let Γ be a lattice in $\mathrm{PSL}(2, \mathbb{R})$ and assume that the stabilizer Γ_∞ of ∞ in Γ is non trivial. Then, for $C > 0$ large enough, the* *following holds. If* z *and* γz *are such that* $\operatorname{Im} z > C$ *and* $\operatorname{Im} \gamma z > C$ *for* $\gamma \in \Gamma$, *then* $\gamma \in \Gamma_\infty$.

Proof By the previous proposition, Γ_∞ is a (discrete) non trivial subgroup of the group

$$\left\{ \begin{pmatrix} 1 & x \\ 0 & 1 \end{pmatrix} \mid x \in \mathbb{R} \right\}.$$

Suppose, by contradiction, that there are sequences $z_n \in \mathbf{H}^2$, and $\gamma_n \in \Gamma$, $\gamma_n \notin \Gamma_\infty$, such that $\operatorname{Im} z_n \geq n$ and $\operatorname{Im} \gamma_n z_n \geq n$. Write

$$\gamma_n = \begin{pmatrix} a_n & b_n \\ c_n & d_n \end{pmatrix}.$$

Then,

$$n \leq \operatorname{Im} \gamma_n z_n = \frac{\operatorname{Im} z_n}{|c_n z_n + d_n|^2} \leq \frac{\operatorname{Im} z_n}{|c_n|^2 |\operatorname{Im} z_n|^2} \leq \frac{1}{|c_n|^2 n}.$$

Hence, $|c_n| \leq 1/n$ and so $c_n \to 0$. Observe that $c_n \neq 0$, since $\gamma_n \notin \Gamma_\infty$. Therefore, by Gaussian elimination, γ_n may be written in the form

$$\gamma_n = \begin{pmatrix} 1 & x_n \\ 0 & 1 \end{pmatrix} \begin{pmatrix} 1 & 0 \\ c_n & 1 \end{pmatrix} \begin{pmatrix} 1 & y_n \\ 0 & 1 \end{pmatrix}.$$

Mutiplying from both sides by suitable matrices from Γ_∞, we obtain a sequence of matrices γ_n' in Γ of the form

$$\gamma_n' = \begin{pmatrix} 1 & x_n' \\ 0 & 1 \end{pmatrix} \begin{pmatrix} 1 & 0 \\ c_n & 1 \end{pmatrix} \begin{pmatrix} 1 & y_n' \\ 0 & 1 \end{pmatrix},$$

with bounded sequences x_n' and y_n'. Passing to a subsequence if necessary, we may assume that x_n' and y_n' converge. As $c_n \to 0$, the sequence γ_n' converges to a matrix from Γ_∞. Since Γ is discrete, we have eventually $c_n = 0$. This is a contradiction. $\qquad\square$

2.19 Theorem. *Let* Γ *be a lattice in* $\mathrm{PSL}(2, \mathbb{R})$ *containing a parabolic element with fixed point* $x \in \mathbb{R} \cup \{\infty\}$. *Let* D *be a Dirichlet region for* Γ. *Then* x *is a vertex at infinity for* δD *for some* $\delta \in \Gamma$. *Moreover, there is an open disc* Ω *in* $\mathbf{H}^2$ *tangent to the real line at* x *with the following properties:*
(i) $\gamma \Omega = \Omega$ *for all* $\gamma \in \Gamma_x$, *the stabilizer of* x *in* Γ;
(ii) $\gamma \Omega \cap \Omega = \emptyset$ *for all* $\gamma \in \Gamma, \gamma \notin \Gamma_x$.

Proof Observe that, if $g \in \mathrm{SL}(2, \mathbb{R})$ is such that $g \cdot \infty = x \in \mathbb{R}$, then g maps the set $\{z \in \mathbf{H}^2 \mid \operatorname{Im} z > C\}$ to an open disc tangent to the real line at x. We may assume that $x = \infty$. Let $C > 0$ be as in the previous lemma, and set

$$\Omega = \{z \in \mathbf{H}^2 | \operatorname{Im} z > C\}.$$

The second claim follows from that lemma.

As to the first claim, observe that some translate δD of D has to contain a Dirichlet region (a "cylinder") for the action of Γ_x on Ω. This proves that $x = \infty$ is a vertex at infinity for δD. $\qquad\square$

With the above notation, the theorem shows that the canonical map from the "cylinder" $\Gamma_x \backslash \Omega$ to $\Gamma \backslash \mathbf{H}^2$ is one-to-one.

A *cusp* of Γ is a Γ-orbit of vertices at infinity for a Dirichlet domain D. It follows from the above that there is a bijection between the cusps of Γ and the conjugacy classes of maximal parabolic subgroups of Γ. (A parabolic subgroup of Γ is the cyclic subgroup generated by a parabolic element in Γ.)

The following is an immediate consequence of Propositions 2.15 and 2.17 and the above theorem.

2.20 Corollary. *A lattice* Γ *in* $\mathrm{SL}(2,\mathbb{R})$ *is cocompact if and only if it has no parabolic elements.*

The following remark concerning the decomposition of the Dirichlet region of a non-cocompact lattice into a compact set and finitely many cusps will play an important rôle in Chapter IV.

2.21 Remark. Let Γ be a non-cocompact lattice in $G = \mathrm{SL}(2,\mathbb{R})$, and let D be a Dirichlet domain for the action of Γ on $\mathbf{H}^2$. In view of Siegel's theorem (Theorem 2.14), D has only finitely many vertices. Let $x_1, \ldots, x_n$ be the vertices at infinity of D. Then $x_i = g_i \infty$ for some $g_i \in \mathrm{SL}(2,\mathbb{R})$, $i = 1, \ldots, n$. For $c > 0$, let Ω_c be the subset

$$\Omega_c = \{z \in \mathbf{H}^2 | \operatorname{Im} z > c\}.$$

of $\mathbf{H}^2$. It is clear that, for suitable $c > 0$, D decomposes as a disjoint union

$$D = \bigcup_{k=1}^{n} (g_k \Omega_c \cap D) \cup D_0,$$

where D_0 is a subset with compact closure in $\mathbf{H}^2$.

§3 The Geodesic Flow of Compact Riemann Surfaces

The Geodesic Flow on $T^1 \mathbf{H}^2$

Since $\mathrm{PSL}(2, \mathbb{R})$ acts by isometries on $\mathbf{H}^2$, it acts on the unit tangent bundle

$$T^1 \mathbf{H}^2 = \{(z, \zeta) \in T\mathbf{H}^2 \mid \|\zeta\|_z = 1\}$$

of $\mathbf{H}^2$ by

$$g \cdot (z, \zeta) = (g \cdot z, D_z g(\zeta)) = \left(g \cdot z, \frac{1}{(cz + d)^2} \zeta \right)$$

for

$$z \in \mathbf{H}^2, \, \zeta \in T_z \mathbf{H}^2, \, g = \pm \begin{pmatrix} a & b \\ c & d \end{pmatrix}.$$

3.1 Lemma. *The action of* $\mathrm{PSL}(2, \mathbb{R})$ *on the unit tangent bundle* $T^1 \mathbf{H}^2$ *is transitive and free, that is, all isotropy groups are trivial. In this way,* $T^1 \mathbf{H}$ *may be identified with* $\mathrm{PSL}(2, \mathbb{R})$.

Proof Take $z_0 = i$ and ζ_0 the unit tangent vector at z_0 pointing in the positive direction of the imaginary axis. Let $(z, \zeta) \in T^1 \mathbf{H}^2$. Let L be the positive imaginary half axis, and let σ be the unique geodesic determined by (z, ζ). Now choose $g \in \mathrm{PSL}(2, \mathbb{R})$ with $g \cdot L = \sigma$ and $g(z_0) = z$. Since transformations from $\mathrm{PSL}(2, \mathbb{R})$ preserve orientation, it is clear that g is unique if one imposes, in addition, the condition that $D_{z_0} g(\zeta_0) = \zeta$. $\square$

Let $z_0 = i$ and ζ_0 be the unit tangent vector as in the above proof. The previous lemma sets up a bijection

$$T^1 \mathbf{H}^2 \to \mathrm{PSL}(2, \mathbb{R}), \, (z, \zeta) \mapsto g_{z\zeta},$$

where $g_{z\zeta}$ is the unique element g in $\mathrm{PSL}(2, \mathbb{R})$ with $g \cdot (z_0, \zeta_0) = (z, \zeta)$. The uniqueness of the element $g_{z\zeta}$ shows that this diffeomorphism intertwines the action of $\mathrm{PSL}(2, \mathbb{R})$ on $T^1 \mathbf{H}^2$ with the left multiplication in the group, that is

$$g \cdot (z, \zeta) \mapsto g \cdot g_{z\zeta}, \qquad \forall g \in \mathrm{PSL}(2, \mathbb{R}). \tag{$*$}$$

3.2 Proposition. *The geodesic flow on* $T^1 \mathbf{H}^2$ *corresponds to the flow on the group* $\mathrm{PSL}(2, \mathbb{R})$ *given by the right translations*

$$g \mapsto g \, g_t,$$

where

$$g_t = \begin{pmatrix} e^{t/2} & 0 \\ 0 & e^{-t/2} \end{pmatrix},$$

for all $t \in \mathbb{R}$.

Proof   From the description of the geodesics in $\mathbf{H}^2$ it is clear that

$$\varphi_t(z_0, \zeta_0) = g_t \cdot (z_0, \zeta_0).$$

Now, for arbitrary $(z, \zeta) \in T^1 \mathbf{H}^2$, let $g_{z\zeta} \in \mathrm{PSL}(2, \mathbb{R})$ be such that

$$(z, \zeta) = g_{z\zeta} \cdot (z_0, \zeta_0).$$

As $g_{z\zeta}$ is an isometry,

$$\varphi_t(z, \zeta) = \varphi_t(g_{z\zeta} \cdot (z_0, \zeta_0)) = g_{z\zeta} \cdot \varphi_t(z_0, \zeta_0) = g_{z\zeta} \cdot g_t(z_0, \zeta_0).$$

$\square$

Recall that the measure

$$d\mu := \frac{dx\, dy}{y^2}$$

on $\mathbf{H}^2$, introduced in Lemma 2.10, is invariant under the action of $\mathrm{SL}(2, \mathbb{R})$. Now, introduce local coordinates (z, ϑ) on $T^1 \mathbf{H}^2$, $z \in \mathbf{H}^2$, $\vartheta \in [0, 2\pi[$. Let λ be the measure on $T^1 \mathbf{H}^2$ given locally by

$$d\lambda = d\mu\,(z)d\vartheta,$$

where μ is the invariant measure on $\mathbf{H}^2$. A computation similar to that in Lemma 2.10 shows that λ is preserved by the action of $\mathrm{PSL}(2, \mathbb{R})$ (see next exercise). Observe that, under the identification

$$T^1 \mathbf{H}^2 \cong \mathrm{PSL}(2, \mathbb{R}),$$

λ induces a left Haar measure on $\mathrm{PSL}(2, \mathbb{R})$, which is also right invariant as $\mathrm{PSL}(2, \mathbb{R})$ is unimodular.

3.3 Exercise.   Show that the above measure λ on $T^1 \mathbf{H}^2$ is preserved by the action of $\mathrm{PSL}(2, \mathbb{R})$.

The Geodesic Flow Associated to a Compact Riemann Surface (of genus $g \geq 2$)

Let Σ be a compact Riemann surface of genus ≥ 2. Then Σ is covered by the hyperbolic plane $\mathbf{H}^2$. More precisely, the fundamental group

$\pi_1(\Sigma) = \Gamma$ of Σ acts freely and discontinously on $\mathbf{H}^2$, by deck transformations. So, Γ may be identified with a discrete (cocompact) subgroup of $\mathrm{PSL}(2, \mathbb{R})$ and Σ with the orbit space $\Gamma \setminus \mathbf{H}^2$. With respect to the structure inherited from $\mathbf{H}^2$, Σ is a Riemannian manifold with constant negative curvature -1. (For all this, see [FK], Chap. IV.)

3.4 Proposition. *The identification* $T^1 \mathbf{H}^2 \cong \mathrm{PSL}(2, \mathbb{R})$ *induces an identification*

$$T^1 (\Gamma \setminus \mathbf{H}^2) \cong \Gamma \setminus \mathrm{PSL}(2, \mathbb{R})$$

of the unit tangent bundle of $\Sigma \cong \Gamma \setminus \mathbf{H}^2$ *with the orbit space* $\Gamma \setminus \mathrm{PSL}(2, \mathbb{R})$. *The geodesic flow on* $T^1(\Sigma)$ *corresponds to the flow*

$$\Gamma \setminus \mathrm{PSL}(2, \mathbb{R}) \to \Gamma \setminus \mathrm{PSL}(2, \mathbb{R})$$

$$\Gamma g \mapsto \Gamma g\, g_t,$$

where

$$g_t := \begin{pmatrix} e^{t/2} & 0 \\ 0 & e^{-t/2} \end{pmatrix},$$

for all $t \in \mathbb{R}$.

Proof Since, by $(*)$ above, $(z, \zeta) \mapsto g_{z\zeta}$ intertwines the actions of $\mathrm{PSL}(2, \mathbb{R})$, this follows from Proposition 3.2. $\qquad\square$

The measure $d\lambda = d\mu(z)d\vartheta$ on $T^1\mathbf{H}^2$ defined above induces a measure $\tilde{\lambda}$ on

$$T^1(\Gamma \setminus \mathbf{H}^2) = \Gamma \setminus T^1\mathbf{H}^2,$$

determined by the formula

$$\int_{T^1\mathbf{H}^2} f d\lambda = \int_{\Gamma \setminus T^1\mathbf{H}^2} \sum_{\gamma \in \Gamma} f(\gamma x) d\tilde{\lambda}(x)$$

for, say, $f \in C_c(T^1\mathbf{H}^2)$, the continuous functions on $T^1\mathbf{H}^2$ with compact support.

Since $T^1(\Gamma \setminus \mathbf{H}^2) = T^1(\Sigma)$ is compact, it has finite measure and the geodesic flow defines a measure preserving action on a finite measure space.

The arguments to follow do not depend on the fact that Γ is cocompact in $\mathrm{PSL}(2, \mathbb{R})$ and acts freely on $\mathbf{H}^2$. So, we shall assume that Γ is an *arbitrary* lattice in $\mathrm{PSL}(2, \mathbb{R})$. The corresponding geodesic flow is defined to be the flow on $\Gamma \setminus \mathrm{PSL}(2, \mathbb{R})$ given as above by right translations by g_t for all $t \in \mathbb{R}$.

In order to prove the ergodicity of the geodesic flow, we have to study the restriction to the subgroup

$$A := \left\{ \begin{pmatrix} a & 0 \\ 0 & a^{-1} \end{pmatrix} \mid a > 0 \right\}$$

of the unitary representation $g \to T_g$ of $\mathrm{SL}(2, \mathbb{R})$ on the Hilbert space

$$\mathcal{H} = L^2 \left(\Gamma \setminus \mathrm{PSL}(2, \mathbb{R}) \right)$$

given by the right translations

$$T_g f(\Gamma x) = f(\Gamma x\, g), \quad \forall x \in \mathrm{PSL}(2, \mathbb{R})) \quad \text{and} \quad f \in \mathcal{H},$$

for $g \in \mathrm{SL}(2, \mathbb{R})$. Consider the following subgroups

$$N = \left\{ \begin{pmatrix} 1 & x \\ 0 & 1 \end{pmatrix} \mid x \in \mathbb{R} \right\}$$

$$N^- = \left\{ \begin{pmatrix} 1 & 0 \\ x & 1 \end{pmatrix} \mid x \in \mathbb{R} \right\}.$$

of $\mathrm{SL}(2, \mathbb{R})$. Observe that N and N^- are normalized by A. The following is a trivial, but important fact.

3.5 Lemma. *For*

$$g = \begin{pmatrix} a & 0 \\ 0 & a^{-1} \end{pmatrix} \in A$$

and $h \in N$ *(respectively* $h \in N^-$ *), one has*

$$\lim_{n \to +\infty} g^n\, h\, g^{-n} = e \quad \text{if } a < 1,$$

respectively

$$\lim_{n \to +\infty} g^n\, h\, g^{-n} = e \quad \text{if } a > 1.$$

Proof One has

$$\begin{pmatrix} a & 0 \\ 0 & a^{-1} \end{pmatrix} \begin{pmatrix} 1 & x \\ 0 & 1 \end{pmatrix} \begin{pmatrix} a & 0 \\ 0 & a^{-1} \end{pmatrix}^{-1} = \begin{pmatrix} 1 & a^2 x \\ 0 & 1 \end{pmatrix}.$$

So,

$$\begin{pmatrix} a & 0 \\ 0 & a^{-1} \end{pmatrix}^n \begin{pmatrix} 1 & x \\ 0 & 1 \end{pmatrix} \begin{pmatrix} a & 0 \\ 0 & a^{-1} \end{pmatrix}^{-n} = \begin{pmatrix} 1 & a^{2n} x \\ 0 & 1 \end{pmatrix},$$

for all $n \in \mathbb{Z}$. Similarly, for all $n \in \mathbb{Z}$,

$$\begin{pmatrix} a & 0 \\ 0 & a^{-1} \end{pmatrix}^n \begin{pmatrix} 1 & 0 \\ x & 1 \end{pmatrix} \begin{pmatrix} a & 0 \\ 0 & a^{-1} \end{pmatrix}^{-n} = \begin{pmatrix} 1 & 0 \\ a^{-2n} x & 1 \end{pmatrix}.$$

$\square$

The following crucial fact is also known as the Mautner phenomenon. The elementary proof we give is due to Green ([AG], Appendix II). Mautner's original proof ([Mau], Lemma 7) used the representation theory of the so-called $ax + b$–group.

3.6 Lemma (Mautner Lemma). *Let G be a locally compact group, and let $(\pi, \mathcal{H}_\pi)$ be a (strongly continuous) unitary representation of G on some Hilbert space $\mathcal{H}_\pi$. Assume that $g, h \in G$ satisfy*

$$\lim_{n \to \infty} g^n h g^{-n} = e,$$

e being the group unit. Then any vector $\xi \in \mathcal{H}_\pi$ that is fixed by g is also fixed by h.

Proof We have, since ξ is fixed by $\pi(g)$ and since $\pi(g)$ is unitary, for any $n \in \mathbb{N}$,

$$\| \pi(h)\xi - \xi \| = \| \pi(h)\pi(g^{-n})\xi - \pi(g^{-n})\xi \|$$
$$= \| \pi(g^n)\pi(h)\pi(g^{-n})\xi - \xi \|.$$

As $g^n h g^{-n} \to e$ and π is strongly continuous, it follows that

$$\pi(h)\xi = \xi.$$

□

3.7 Corollary. *Let $(\pi, \mathcal{H}_\pi)$ be a unitary representation of $\mathrm{SL}(2, \mathbb{R})$ in some Hilbert space $\mathcal{H}_\pi$. Let $\xi \in \mathcal{H}_\pi$ be such that $\pi(A)\xi = \xi$. Then ξ is invariant under $\mathrm{SL}(2, \mathbb{R})$.*

Proof By the previous lemmas

$$\pi(N)\xi = \xi \quad \text{and} \quad \pi(N^-)\xi = \xi.$$

Since $N \cup N^-$ generates $\mathrm{SL}(2, \mathbb{R})$, the result follows. □

3.8 Corollary. *The geodesic flow of $\Gamma \backslash \mathrm{PSL}(2, \mathbb{R})$ is ergodic, for any lattice Γ in $\mathrm{PSL}(2, \mathbb{R})$.*

Proof Take π to be the unitary representation $g \mapsto T_g$ of $\mathrm{SL}(2, \mathbb{R})$ defined by right translations on $L^2(\Gamma \backslash \mathrm{PSL}(2, \mathbb{R}))$. The previous corollary shows that any A-invariant function in $L^2(\Gamma \backslash \mathrm{PSL}(2, \mathbb{R}))$ is invariant under $\mathrm{SL}(2, \mathbb{R})$ and, hence, constant almost everywhere. □

3.9 Remark. As an application of the Howe–Moore vanishing theorem, we shall prove in Chap. III, Corollary 2.3, that the above flow is strongly mixing.

§4 The Geodesic Flow of Riemannian Locally Symmmetric Spaces

We shall describe the geodesic flow of Riemannian locally symmmetric spaces, generalizing in this way the results about surfaces from the previous section.

Riemannian Symmetric Spaces

4.1 Definition. A Riemannian manifold (M, Q) is called a *Riemannian globally symmetric space* if, for every $p \in M$, there exists an involutive isometry s_p with p as isolated fixed point. The isometry s_p is the called the *symmetry* around p.

4.2 Remark. The symmetry around p coincides in a neighbourhood of p with the geodesic symmetry $\gamma(t) \mapsto \gamma(-t)$, where γ is a geodesic arc with $\gamma(0) = p$ (see [Hel], Chap. IV, Lemma 3.1). Here is the argument.

Let $A = ds_p$ be the differential of s_p in p. Then, as p is fixed under s_p and as s_p is an involution, the tangent space $T_p M$ is invariant under A and $A^2 = \mathrm{id}$. Write

$$T_p M = V^+ \oplus V^-,$$

where $V^\pm$ is the ± 1-eigenspace of A. Suppose that there exists $v \in V^+, v \neq 0$. Let γ_v be the geodesic through p with $\dot{\gamma}_v(0) = v$. Then, $\tilde{\gamma} = s_p \circ \gamma_v$ is a geodesic through p with $\dot{\tilde{\gamma}}(0) = Av = v$. Hence, $s_p \circ \gamma_v = \gamma_v$. This contradicts the fact that p is an isolated fixed point of s_p. Hence, $V^+ = 0$, that is, $ds_p = -\mathrm{id}$, and, as s_p is an isometry, the claim follows.

4.3 Examples. Here are some examples of Riemannian globally symmetric spaces:
(i) The Euclidean space $\mathbb{R}^n$: s_p is the reflection about $p \in \mathbb{R}^n$.
(ii) The sphere S^2 : s_p is the rotation around the axis through p by the angle π.
(iii) The hyperbolic plane $\mathbf{H}^2$: for $z = i$ take $s_i(z) := -1/z$. For arbitrary w, let g be a Möbius transformation with $gi = w$ and take $s_w := g \circ s_i \circ g^{-1}$.

Riemannian symmetric pairs – to be defined now – give rise to Riemannian symmetric spaces.

4.4 Definition. Let G be a connected Lie group with Lie algebra $\mathfrak{g}$. Let K be a closed subgroup of G such that $\mathrm{Ad}(K)$ is a compact subgroup of $\mathrm{Aut}(\mathfrak{g})$. (Ad denotes the adjoint representation of G.) The pair (G, K) is called a *Riemannian symmetric pair*, if there is an involutive automorphism σ of G such that

$$(G_\sigma)_0 \subseteq K \subseteq G_\sigma,$$

where $G_\sigma = \{g \in G \mid \sigma(g) = g\}$ and $(G_\sigma)_0$ is the identity component of σ.

4.5 Example. Let $G = \mathrm{SL}(n, \mathbb{R})$ and $K = \mathrm{SO}(n)$. Let σ be the involution on G given by

$$\sigma(A) = (A^{-1})^{\mathrm{t}}.$$

Then $\mathrm{SO}(n) = G_\sigma$ and $(\mathrm{SL}(n, \mathbb{R}), \mathrm{SO}(n))$ is a Riemannian symmetric pair.

4.6 Theorem. *Let (G, K) be a Riemannian symmetric pair with involution σ. Then there exists a G-invariant Riemannian structure Q on G/K, and in each such structure Q, the homogeneous space G/K is a globally Riemannian symmetric space, with symmetry s_{gK} at gK defined by*

$$s_{gK}(hK) := g\sigma(g^{-1}h)K.$$

Proof Let $\mathfrak{g}$ and $\mathfrak{k}$ be the Lie algebras of G and K, respectively. Let $d\sigma$ and $d\pi$ respectively be the differential of σ and of the canonical projection $\pi : G \to G/K$ at the group unit $e \in G$. Let

$$\mathfrak{p} := \{X \in \mathfrak{g} \mid d\sigma(X) = -X\}.$$

Then $\mathfrak{g} = \mathfrak{k} \oplus \mathfrak{p}$. For $X \in \mathfrak{p}$, $k \in K$ and $t \in \mathbb{R}$, we have

$$\sigma(\exp(\mathrm{Ad}(k)tX)) = k\exp(-tX)k^{-1}.$$

So, $d\sigma(\mathrm{Ad}(k)X) = -\mathrm{Ad}(k)X$. Hence, $\mathfrak{p}$ is $\mathrm{Ad}K$-invariant.
Observe that $\ker d\pi = \mathfrak{k}$. Thus,

$$d\pi : \mathfrak{p} \to T_K(G/K), \quad X \mapsto d\pi(X)$$

is a linear bijection.

Now the group G acts on G/K via $\lambda(g_0)gK = g_0gK$. For $k \in K$, $X \in \mathfrak{p}$ and $t \in \mathbb{R}$, we have

$$\exp(\mathrm{Ad}(k)tX) \cdot K = k\exp(tX)k^{-1} \cdot K = \lambda(k)\exp(tX) \cdot K,$$

and, hence,

$$d\pi(\mathrm{Ad}(k)X) = d\lambda(k)_K \circ d\pi(X). \qquad (*)$$

Here, $d\lambda(k)_K$ denotes the differential of the diffeomorphism $\lambda(k)$ at the coset K. Since $\mathfrak{p}$ is K–invariant and $\mathrm{Ad}(K)$ is compact, there exists an $\mathrm{Ad}(K)$–invariant inner product B on $\mathfrak{p}$. Then the inner product $Q_K := B \circ (d\pi)^{-1}$ on $T_K(G/K)$ is $d\lambda(K)_K$–invariant by $(*)$. Now (identifying $\mathfrak{p}$ and $T_K(G/K)$) define a bilinear form on $T_{gK}(G/K)$ by

$$Q_{gK}(d\lambda(g)_K X, d\lambda(g)_K Y) := Q_K(X,Y), \quad \forall X,Y \in \mathfrak{p}.$$

Since Q_K is K–invariant, Q_{gK} is well-defined and $gK \mapsto Q_{gK}$ is a G–invariant Riemannian structure on G/K.

Clearly, each G-invariant Riemannian structure on G/K arises in this fashion from an $\mathrm{Ad}(K)$-invariant inner product on $\mathfrak{p}$.

It remains to define the symmetries. The mappings s_{gK} as defined in the statement of the theorem are involutive diffeomorphisms. Since $\lambda(g)$ is an isometry for each $g \in G$, it suffices to show that s_K is an isometry. So, let $X,Y \in T_{gK}(G/K)$. Then $X_0 = d\lambda(g^{-1})_{gK}(X)$, $Y_0 = d\lambda(g^{-1})_{gK}(Y) \in T_K(G/K)$. Next, observe that for $g,h \in G$

$$s_K \circ \lambda(g)(hK) = \sigma(gh)K = \sigma(g)\sigma(h)K = \lambda(\sigma(g)) \circ s_K(hK). \qquad (**)$$

Moreover, $(ds_K)_K = -\mathrm{id}$. It follows that

$$Q_{\sigma(g)K}((ds_K)_{gK}(X), (ds_K)_{gK}(Y)) =$$
$$= Q_{\sigma(g)K}((ds_K)_{gK}(d\lambda(g)_K(X_0)), (ds_K)_{gK}(d\lambda(g)_K(Y_0)))$$
$$= Q_K((ds_K)_K(X_0), (ds_K)_K(Y_0)) = Q_K(X_0, Y_0) = Q_{gK}(X,Y).$$

Hence, s_K is an isometry, as stated. $\qquad \square$

4.7 Remark. Conversely, all Riemannian symmetric spaces arise from Riemannian symmetric pairs in the way described above. Here is the precise statement.

Let M be a Riemannian symmetric space, $p \in M$, and s_p the symmetry around p. Let G be the identity component of the group of all isometries on M and K the stabilizer of p in G. Then (G,K) is a Riemannian symmetric pair with respect to the involutive automorphism $\sigma(g) := s_p g s_p$ and G/K is diffeomorphic to M via $gK \mapsto g \cdot p$.

For the proof of this fact which lies beyond the scope of these notes, we refer to [Hel], Chap. IV, Theorem 3.3.

Locally Symmetric Riemannian Spaces

4.8 Definition. An analytic Riemannian manifold is called a *Riemannian locally symmetric space* if, for every $p \in M$, the geodesic symmetry $\gamma(t) \mapsto \gamma(-t)$ (which is defined for sufficiently small $t \in \mathbb{R}$) is a local isometry.

Riemannian locally symmetric spaces arise as follows. Let N be a globally Riemannian symmetric space which covers a manifold M. One defines the Riemannian metric Q on M to be locally the same as the Riemannian metric $\widetilde{Q}$ on N. More precisely, let $\pi : N \to M$ be the covering mapping. Given $p \in M$, choose $\tilde{p} \in \pi^{-1}(p)$. For $X_p, Y_p \in T_p(M)$, define

$$Q_p(X_p, Y_p) = \widetilde{Q}_{\tilde{p}}(d\pi_p^{-1}(X_p), d\pi_p^{-1}(Y_p)).$$

This definition does not depend on the choice of $\tilde{p} \in \pi^{-1}(p)$. Indeed, if $\tilde{q} \in \pi^{-1}(p)$, then there exists an isometry φ of N with $\varphi(\tilde{p}) = \tilde{q}$, since the group of isometries of N acts transitively. This proves the claim. With respect to this Riemannian structure, the covering mapping π is a local isometry. Hence, the geodesics on M are the projections of those of N, and the geodesic symmetry is a local isometry. Conversely, the universal cover of a locally symmetric space can be made into a globally symmetric space, such that the covering mapping is a local isometry. For all this, see [Hel], Chapter IV.

The fundamental group of a locally symmetric space M is a discrete subgroup of the group G of isometries of the universal covering $\widetilde{M}$, acting properly discontinuously and freely on $\widetilde{M}$.

We recall some standard facts about proper group actions. Let G be a locally compact group acting on a locally compact space M. The action of G is said to be *proper* if, for any compact subsets K, L of M, the set

$$\{g \in G \mid gK \cap L \neq \emptyset\}$$

is compact. (If G is discrete, one usually speaks of a properly discontinuous action; see Definition 2.2)

Let Γ be a discrete group with a properly discontinuous action on M. Clearly, for any $x \in M$, the stabilizer Γ_x of x is finite. It is straightforward to verify that the following holds:

(i) Any $x \in M$ has a neighbourhood U such that

$$\{g \in \Gamma \mid gU \cap U \neq \emptyset\} = \Gamma_x.$$

(ii) If $x, y \in M$ are not in the same orbit, then there are neighbourhoods U, V of x, y such that $U \cap \Gamma \cdot V = \emptyset$.

Observe that (ii) says that the orbit space M/Γ is a Hausdorff space.

Recall that a group acts freely on M if each point of M has trivial stabilizer.

4.9 Lemma. *Let Γ be a discrete group acting properly discontinuously and freely on the manifold $\widetilde{M}$. Then there is a unique C^∞-structure on $M = \widetilde{M}/\Gamma$ (compatible with the quotient topology) such that the projection $\pi : \widetilde{M} \to M$ is a locally diffeomorphic covering.*

Proof As mentioned above, M is a Hausdorff space. Furthermore, the map π is obviously open with respect to the quotient topology and M has a countable basis of the topology. Now, for each $x \in \widetilde{M}$, $\Gamma_x = e$ and we find a neighbourhood $\widetilde{U}$ such that $h\widetilde{U} \cap \widetilde{U} = \emptyset$ if $h \neq e$, by (i) above. Then π, restricted to $\widetilde{U}$, is a homeomorphism onto an open neighbourhood U of $\Gamma \cdot x \in M$. We may assume without loss of generality that $\widetilde{U}$ is the domain for a chart $\widetilde{\varphi}$ on $\widetilde{M}$ around x. Then

$$\varphi : U \to \mathbb{R}^n, \quad u \mapsto \widetilde{\varphi} \circ (\pi|_{\widetilde{U}})^{-1}(u)$$

is a chart on M around $\Gamma \cdot x$, where $n = \dim \widetilde{M}$. One easily verifies that these charts define a C^∞-structure on M satisfying all requirements. $\square$

The following proposition generalizes the fact, proved in Lemma 2.3, that a discrete of $\mathrm{SL}(2, \mathbb{R})$ acts properly discontinuously on

$$\mathbf{H}^2 \cong \mathrm{SL}(2, \mathbb{R})/\mathrm{SO}(2).$$

4.10 Proposition. *Let G be a locally compact group, and let K be a compact subgroup of G. Then any closed subgroup H acts properly on the homogeneous space $M = G/K$.*

Proof Let L_1, L_2 be compact subsets of M. Then, since K is compact, $L_1' := \pi^{-1}(L_1)$ and $L_2' := \pi^{-1}(L_2)$ are compact subsets of G, where $\pi : G \to M$ is the canonical projection. Clearly, $gL_1 \cap L_2 \neq \emptyset$ if and only if $g \in L_2' L_1'^{-1}$. This proves the claim. $\square$

The Geodesic Flow of Riemannian Symmetric Spaces

We shall be interested in Riemannian symmetric spaces with a noncompact semisimple Lie group of isometries. References for the generalities on semisimple groups we are going to recall are [Hel] or [Va2].

Let G be a *non-compact* connected semisimple Lie group with finite centre and with Lie algebra $\mathfrak{g}$. Let

$$\kappa(X, Y) = \operatorname{trace}(\operatorname{ad} X \operatorname{ad} Y), \qquad X, Y \in \mathfrak{g}$$

be the Killing form of $\mathfrak{g}$. There exists a *Cartan involution* θ on $\mathfrak{g}$. This is an involutive automorphism θ of $\mathfrak{g}$ such that $\kappa|_{\mathfrak{k}}$ is negative definite and $\kappa|_{\mathfrak{p}}$ is positive definite, where

$$\mathfrak{k} := \{X \in \mathfrak{g} | \ \theta(X) = X\},$$
$$\mathfrak{p} := \{X \in \mathfrak{g} | \ \theta(X) = -X\},$$

are the 1 and -1 eigenspaces of θ. The decomposition $\mathfrak{g} = \mathfrak{k} \oplus \mathfrak{p}$ is the *Cartan decomposition* of $\mathfrak{g}$.

There is also a global decomposition. The automorphism θ of $\mathfrak{g}$ extends to an involutive automorphism $\tilde{\theta}$ of G (that is, θ is the differential at e of $\tilde{\theta}$). Let $K = G_{\tilde{\theta}}$, the subgroup of all $\tilde{\theta}$-fixed elements in G. Then K is a connected, maximal compact subgroup of G and there is a diffeomorphism

$$\Phi : K \times \mathfrak{p} \to G, \quad (k, X) \mapsto k \exp(X),$$

called the *Cartan decomposition* of G. One has

$$\tilde{\theta}(k \exp(X)) = k \exp(-X) \qquad k \in K, X \in \mathfrak{p}.$$

4.11 Example. Let $G = \operatorname{SL}(n, \mathbb{R})$. Its Lie algebra is

$$\mathfrak{sl}(n, \mathbb{R}) = \{X \in M(n, \mathbb{R}) | \operatorname{trace} X = 0\}.$$

with Killing form

$$\kappa(X, Y) = 2n \operatorname{trace}(XY).$$

The involution $\theta(X) = (-X)^t$ on $\mathfrak{sl}(n, \mathbb{R})$ is a Cartan involution with 1 and -1 eigenspaces

$$\mathfrak{k} = \mathfrak{so}(n, \mathbb{R}) = \{X \in \mathfrak{sl}(n, \mathbb{R}) | X^t = -X\}$$

$$\mathfrak{p} = \{X \in \mathfrak{sl}(n, \mathbb{R}) | X^t = X\}.$$

The extension of θ to G is the automorphism $\tilde{\theta}(A) = (A^{-1})^t$ from Example 4.5. The Cartan decomposition is the usual polar decomposition $\operatorname{SL}(n, \mathbb{R}) = \operatorname{SO}(n)P$, where P is the set of all positive definite symmetric matrices.

We return to our general setting, with G, K as above. As seen in Theorem 4.6, $M := G/K$ carries a G-invariant Riemannian structure Q

and (M, Q) is a globally symmetric space with the symmetry

$$s_{gK}(hK) = g\theta(hg^{-1})K$$

at gK. We shall identify the tangent space $T_K M$ with $\mathfrak{p}$, by means of the isomorphism

$$d\pi : \mathfrak{p} \to T_K(G/K),$$

where $d\pi$ is the differential at e of the projection $\pi : G \to G/K$. The action of K on $T_K(G/K)$ corresponds to the action of K on $\mathfrak{p}$ by the adjoint representation Ad. The inner product Q_K on $T_K(G/K)$ is then identified with an $\mathrm{Ad}(K)$-invariant inner product on $\mathfrak{p}$.

We now describe the geodesics in M. Recall that a Riemannian manifold M is *complete* if M, with the induced metric d (as defined in 1.3), is a complete metric space. By the Hopf–Rinow theorem (see [Hel], Chap. I, §10), this is equivalent to the following fact: for every $p \in M$ and every v in the tangent space $T_p(M)$, the unique geodesic γ_v with $\gamma_v(0) = p$, $\dot{\gamma}_v(0) = v$ is globally defined. In this case, one has $\gamma_{tv}(s) = \gamma_v(st)$ for all $s, t \in \mathbb{R}$, and any two points $p, q \in M$ are connected by a geodesic curve γ_v with $\|v\| = 1$.

In the sequel, we denote by o the base point $K \in G/K$.

4.12 Proposition. *For $v \in \mathfrak{p}$, the unique geodesic γ_v through the point $o := K$ in $M = G/K$ with $\dot{\gamma}_v(0) = v$ is*

$$t \mapsto \gamma_v(t) = \exp(tv) \cdot o. \tag{1}$$

In particular, M is a complete Riemannian symmetric space, and, for a unit vector $v \in \mathfrak{p}$,

$$d(o, \exp(tv) \cdot o) = |t|.$$

Proof Set $\gamma(t) = \exp(tv) \cdot o$. Clearly, $\gamma(0) = o$ and $\dot{\gamma}(0) = v$. We have to show that γ is a geodesic. Let $p_t = \gamma(t)$ be an arbitrary point on γ. Observe that

$$s_{p_t}(\gamma(s)) = \gamma(2t - s).$$

Thus s_{p_t} reflects γ around p_t. Fix $t_0 < t_1$ with $t_1 - t_0$ sufficiently small. Let γ' be the unique geodesic joining p_{t_0} to p_{t_1}. Let $t_2 = (t_1 + t_0)/2$. By the above, $s_{p_{t_2}}(p_{t_1}) = p_{t_0}$. Hence, since $s_{p_{t_2}}$ is an isometry, $s_{p_{t_2}} \circ \gamma'$ is a geodesic joining p_{t_1} to p_{t_0}. Therefore, $s_{p_{t_2}} \circ \gamma'$ and γ' differ only by a continuous change of parameter. Hence, by an elementary fixed point theorem, $s_{p_{t_2}} \circ \gamma'(s) = \gamma'(s)$ for some s. Since p_{t_2} is an isolated fixed point of $s_{p_{t_2}}$, it follows that $\gamma'(s) = p_{t_2}$. So, γ' intersects γ at p_{t_2}. We may repeat the same argument with the intervals $[t_0, t_2]$ and $[t_2, t_1]$ instead of $[t_0, t_1]$ and continue this way. Hence, by continuity, the images of γ' and γ coincide on $[t_0, t_1]$. This shows that γ is a geodesic.

The length of γ on $[0, t]$ is

$$\int_0^t \|\dot\gamma(s)\| \, ds = t\|v\|,$$

since $\|\dot\gamma(s)\| = \|\exp(sv)v\| = \|v\|$. $\qquad\square$

Let

$$\mathfrak{p}_1 := \{v \in \mathfrak{p} | \; Q_K(v, v) = 1\},$$

the unit sphere in $\mathfrak{p}$. Recall that the group G acts by isometries on M via $\lambda(g)hK = g \cdot hK = ghK$ and, hence, on the unit tangent bundle T^1M via

$$g \cdot (p, v) := (g \cdot p, d\lambda(g)_p(v)).$$

For $v \in \mathfrak{p}$, we shall denote the element (o, v) of T^1M by the same letter v. So, for $g \in G$, $g \cdot v$ is the tangent vector

$$g \cdot (o, v) = (g \cdot o, d\lambda(g)_o(v))$$

based at $g \cdot o$.

Observe that

$$T^1M = G \cdot \mathfrak{p}_1 = \{g \cdot v | \; g \in G, v \in \mathfrak{p}_1\}.$$

Indeed, G acts transitively on G/K and $d\lambda(g)_{h \cdot o}$ is a linear bijection of $T_{h \cdot o}M$ onto $T_{gh \cdot o}M$ for all $g, h \in G$. Since G acts by isometries on M, we find for every $(p, v) \in T^1M$, elements $w \in \mathfrak{p}_1$ and $g \in G$ with $(p, v) = g \cdot (o, w)$.

4.13 Corollary. *For every $g \in G$, the geodesic through the point $g \cdot o \in M = G/K$ in the positive direction of $d\lambda(g)_o(v)$, $v \in \mathfrak{p}_1$, is the curve*

$$t \mapsto \gamma(t) = g \exp(tv) \cdot o.$$

Proof This follows from Proposition 4.12 and the fact that g is an isometry of M. $\qquad\square$

We now describe the geodesic flow on G/K.

4.14 Theorem. *The geodesic flow φ_t on $M = G/K$ is given as follows:*

$$t \mapsto \varphi_t(g \cdot v) = g \exp(tv) \cdot v, \qquad v \in \mathfrak{p}_1, g \in G.$$

Proof Consider an arbitrary point $g \cdot v$ in T^1M ($g \in G, v \in \mathfrak{p}_1$). By Corollary 4.13 above, the geodesic in M through $g \cdot v$ is $\gamma(t) = g \exp(tv) \cdot o$. The unit tangent vector in the positive direction at $\gamma(t_0)$, $t_0 \in \mathbb{R}$, is $d\lambda(g \exp(t_0v))_o(v)$. By Proposition 4.12, the distance between $g \cdot o$ and

$g \exp(t_0 v) \cdot o$ is t_0. Hence,

$$\varphi_{t_0}(g \cdot v) = (g \exp(t_0 v) \cdot o, d\lambda(g \exp(t_0 v))_o(v)) = g \exp(t_0 v) \cdot v,$$

as claimed.   □

4.15 Exercise. Let $G = \mathrm{SL}(2, \mathbb{R}), K = \mathrm{SO}(2)$ and $\mathfrak{p}$ the symmetric 2×2 matrices of trace 0. Show that

$$\mathfrak{p} \to T_i \mathbf{H}^2, \quad v \mapsto \dot\gamma_v(0),$$

where $\gamma_v = \exp tv \cdot i$, identifies $T_e G/K \cong T_i \mathbf{H}^2$ with $\mathfrak{p}$. For

$$v_0 = \begin{pmatrix} 1/2\sqrt{2} & 0 \\ 0 & -1/2\sqrt{2} \end{pmatrix} \in \mathfrak{p}_1,$$

identify the corresponding unit tangent vector $\zeta_0 \in T_i \mathbf{H}^2$. Deduce the description of the geodesic flow of $\mathbf{H}^2$ given in Proposition 3.2 from the one given above.

Geodesic Flow for Locally Symmetric Spaces

Let the notation be as above and let Γ be a lattice in G acting freely on $M = G/K$. By Proposition 4.10, Γ acts properly discontinuously on M. Hence, the canonical mapping $\pi : M \to \Gamma \backslash M$ is a covering mapping. Defining the metric Q^Γ to be locally the same as the metric on M (see the remarks after Definition 4.8), $\Gamma \backslash M$ is a Riemannian manifold. The canonical mapping $T^1 M \to T^1(\Gamma \backslash M)$ gives an identification of $T^1(\Gamma \backslash M)$ with the orbit space $\Gamma \backslash T^1 M$. Since the geodesics in $\Gamma \backslash M$ are projections of the geodesics of M, the following is now an immediate consequence of the description of the geodesic flow on M.

4.16 Theorem. *With the above notations, $(\Gamma \backslash M, Q^\Gamma)$ is a Riemannian locally symmetric space. The unit tangent bundle $T^1(\Gamma \backslash M)$ can be identified with the orbit space $\Gamma \backslash T^1 M$. In particular, every element of $\Gamma \backslash T^1 M$ is of the form $\Gamma g \cdot v$ for suitable $g \in G$ and $v \in \mathfrak{p}_1$. The geodesic flow on $\Gamma \backslash M$ is given by*

$$t \mapsto \varphi_t(\Gamma g \cdot v) = \Gamma g \exp(tv) \cdot v, \qquad g \in G, v \in \mathfrak{p}_1.$$

Let dp be the canonical volume form on the Riemannian manifold $M = G/K$ and dv the canonical rotation invariant measure on the unit sphere $\mathfrak{p}_1$.

4.17 Proposition. *The measure $d\mu$ on T^1M, defined locally by $d\mu = dp \cdot dv$, is invariant under the geodesic flow.*

Proof This holds more generally for any Riemannian manifold (see [KH], Proposition 5.3.6). We give a proof in the case $M = G/K$.

Let f be a continuous function on $T^1(G/K)$ with compact support. For $v \in \mathfrak{p}_1$, let f_v be the continuous function on G with compact support defined by $f_v(g) = f(g \cdot v)$ for $g \in G$. Identify dp with the G-invariant measure on G/K. Then, under suitable normalizations,

$$\int_G f_v(g) dg = \int_{G/K} \int_K f_v(gk) dk dp(\dot{g}) \qquad (\dot{g} = gK). \qquad (*)$$

Hence, as $g \mapsto \int_{\mathfrak{p}_1} f_v(g) dv$ is K-invariant,

$$\int_{T^1(G/K)} f \circ \varphi_t d\mu = \int_{G/K} \int_{\mathfrak{p}_1} f(g \exp tv \cdot v) dv dp(\dot{g})$$

$$= \int_{G/K} \int_K \int_{\mathfrak{p}_1} f(g \exp t(\mathrm{Ad}(k)v) \cdot \mathrm{Ad}(k)v) dv dk dp(\dot{g})$$

$$= \int_{G/K} \int_K \int_{\mathfrak{p}_1} f(gk \exp tv \cdot v) dv dk dp(\dot{g})$$

$$= \int_{\mathfrak{p}_1} \left(\int_G f_v(g \exp tv) dg \right) dv$$

$$= \int_{\mathfrak{p}_1} \int_G f_v(g) dg dv,$$

by the right invariance of the left Haar measure dg on G (observe that G is unimodular). Now, using $(*)$,

$$\int_{\mathfrak{p}_1} \int_G f_v(g) dg dv = \int_{T^1(G/K)} f d\mu.$$

$\square$

It follows that, for a lattice Γ in G, the measure $d\mu_\Gamma$ on $T^1(\Gamma \backslash M)$ induced by μ is invariant under the geodesic flow. We want to study the ergodicity of the geodesic flow on $T^1(\Gamma \backslash M)$ with respect to $d\mu_\Gamma$.

The arguments we shall give do not depend on the fact that $\Gamma \backslash M$ is a manifold.

In the sequel, we shall assume that Γ is an arbitrary lattice in the connected non-compact semisimple Lie group G with finite centre and we shall discuss a generalized geodesic flow on $\Gamma \backslash T^1(G/K)$, defined by the formula,

$$\varphi_t(\Gamma g \cdot v) = \Gamma g \exp(tv) \cdot v \qquad \forall g \in G, v \in \mathfrak{p}, \qquad (2)$$

where, as above, K is the set of fixed points of a Cartan involution on G.

Non-ergodicity of the Geodesic Flow in Higher Rank

Recall that, under the identification of the tangent space $T_o(G/K)$ at $o = K$ with $\mathfrak{p}$, the action of K on $T_o(G/K)$ corresponds to the action of K on $\mathfrak{p}$ by the adjoint action Ad. Whether the geodesic flow is ergodic depends, as we now see, on whether K acts transitively on $\mathfrak{p}_1$ or not.

4.18 Lemma.
(i) *Let U be an open subset of $\mathfrak{p}_1$. Then $G \cdot U = \{g \cdot v | g \in G, v \in U\}$ is an open subset of $T^1(G/K)$.*
(ii) *Let A, B be two K-invariant disjoint subsets of $\mathfrak{p}_1$. Then $G \cdot A$ and $G \cdot B$ are disjoint.*

Proof
(i) Let $g \in G$ and $v \in U$. For every neighbourhood V of e in G, $gV \cdot U$ is a neighbourhood of $g \cdot v$.
(ii) Assume that $g_1 \cdot a = g_2 \cdot b$ in $G \cdot A \cap G \cdot B$ for some $g_1, g_2 \in G$, $a \in A$ and $b \in B$. Then $a = g_1^{-1} g_2 \cdot b$. Hence, $g_1^{-1} g_2 \in K$ and, since B is K-invariant, $a \in B$. This is a contradiction, as $A \cap B = \emptyset$. $\qquad \square$

4.19 Corollary. *The geodesic flow on $\Gamma \backslash T^1(G/K)$ is not ergodic if K does not act transitively on $\mathfrak{p}_1$.*

Proof Assume that K does not act transitively on $\mathfrak{p}_1$. Let $\mathrm{Ad}(K)v_1$ and $\mathrm{Ad}(K)v_2$ be two disjoint K-orbits in $\mathfrak{p}_1$. Due to the compactness of K, we find open K-invariant neighbourhoods V_1, V_2 of $K \cdot v_1, K \cdot v_2$ with $V_1 \cap V_2 = \emptyset$. By the above lemma, $G \cdot V_1$ and $G \cdot V_2$ are two open subsets of $T^1(G/K)$ with

$$G \cdot V_1 \cap G \cdot V_2 = \emptyset.$$

Then $\Gamma \backslash G \cdot V_1$ and $\Gamma \backslash G \cdot V_2$ are two open subsets of $\Gamma \backslash T^1(G/K)$ with $\Gamma \backslash (G \cdot V_1) \cap \Gamma \backslash (G \cdot V_1) = \emptyset$. By (2) above, $\Gamma \backslash G \cdot V_1$ and $\Gamma \backslash G \cdot V_2$ are invariant under the geodesic flow. Since the support of the invariant measure under consideration is the whole space $\Gamma \backslash T^1(G/K)$, these open sets have strictly positive measure. Hence, the geodesic flow is not ergodic. $\qquad \square$

4.20 Remark. The action of K on $\mathfrak{p}_1$ is transitive if and only if G acts transitively on $T^1(G/K)$.

We shall see below (Theorem 4.26) that the converse of Corollary 4.19 is true: if K is transitive on $\mathfrak{p}_1$, then the geodesic flow is ergodic.

We now analyze the action of K on $\mathfrak{p}$. To do so, we have to use some standard facts from the theory of semisimple Lie groups.

Fix a maximal abelian subspace $\mathfrak{a}$ of $\mathfrak{p}$. It is known that

$$\mathfrak{p} = \bigcup_{k \in K} \mathrm{Ad}(k)\mathfrak{a}$$

(see [Hel], Chap. V, Lemma 6.3). So, every K orbit in $\mathfrak{p}$ meets $\mathfrak{a}$. (In the case where $G = \mathrm{SL}(n, \mathbb{R})$ and $\mathfrak{a}$ is the space of diagonal matrices in $\mathfrak{g}$, this corresponds to the familiar fact that a symmetric matrix has an orthogonal basis consisting of eigenvectors.)

The bilinear form

$$(X, Y) \mapsto \kappa_\theta(X, Y) := -\kappa(X, \theta Y)$$

is positive definite on $\mathfrak{g}$, where κ is the Killing form and θ the Cartan involution on $\mathfrak{g}$. For H in $\mathfrak{p}$, the linear mapping $\mathrm{ad}\, H$ is symmetric with respect to κ_θ, as is readily verified. Since $\mathfrak{a}$ is abelian, the mappings $\mathrm{ad}\, H$, $H \in \mathfrak{a}$, are simultaneously diagonalizable. For a linear form λ on $\mathfrak{a}$, define

$$\mathfrak{g}^\lambda := \{X \in \mathfrak{g} |\ \mathrm{ad}\, H(X) = \lambda(H)X,\ \forall H \in \mathfrak{a}\}$$

and $\Sigma := \{\lambda \in \mathfrak{a}^* \setminus \{0\} | \mathfrak{g}^\lambda \neq 0\}$, the set of *roots* of $\mathfrak{a}$. One has a corresponding *root space decomposition*

$$\mathfrak{g} = \mathfrak{g}^0 \oplus \sum_{\lambda \in \Sigma} \mathfrak{g}^\lambda.$$

Introduce a vector space order on $\mathfrak{a}^*$. Recall that a vector space order on a finite-dimensional real vector space V is a total order on V, denoted $>$, such that $v + w > 0$ and $\lambda v > 0$ for all $v, w \in V$ with $v > 0, w > 0$ and all positive real numbers λ. Such orders always exist. For instance, identifying V with $\mathbb{R}^n$ by means of a basis, the lexicographic order is a vector space order on V.

Denote by Σ^+ the system of positive roots. The *positive Weyl chamber* is

$$\mathfrak{a}^+ := \{H \in \mathfrak{a} |\ \lambda(H) > 0,\ \forall \lambda \in \Sigma^+\}.$$

The *Weyl group* of Σ is the finite group

$$W = M^*/M \cong \mathrm{Ad}(M^*)|_\mathfrak{a},$$

where

$$M^* = \{k \in K |\ \mathrm{Ad}(k)\mathfrak{a} = \mathfrak{a}\} \quad \text{and} \quad M = \{k \in K |\ \mathrm{Ad}(k)|_\mathfrak{a} = \mathrm{id}_\mathfrak{a}\}$$

are the normalizer and the centralizer of $\mathfrak{a}$ in K, respectively. Let $\overline{\mathfrak{a}^+}$ be the closure of $\mathfrak{a}^+$. Now, any H in $\mathfrak{a}$ is conjugate under the Weyl group W to a unique element from $\overline{\mathfrak{a}^+}$. Thus, the set $\overline{\mathfrak{a}^+}$ parametrizes the K-orbits in $\mathfrak{p}$.

Now, all $\mathrm{Ad}(k), k \in K$, are isometries on $\mathfrak{p}$, with respect to the Killing form κ. Hence, the above discussion shows the following.

4.21 Theorem. *The intersection $\overline{\mathfrak{a}_1^+}$ of the closed positive Weyl chamber $\overline{\mathfrak{a}^+}$ with the unit sphere in $\mathfrak{a}$ is a fundamental domain for the action of K on $\mathfrak{p}_1$, in the sense that, for every $X \in \mathfrak{p}_1$, there is a unique $H \in \overline{\mathfrak{a}_1^+}$ in the K-orbit of X.*

The *rank* of the symmetric space G/K is the dimension of a maximal abelian subspace $\mathfrak{a}$ of $\mathfrak{p}$. This dimension is also called the (real) rank of G.

If $\dim \mathfrak{a} \geq 2$, then $\dim \mathfrak{a}_1^+ \geq 1$. In particular, K cannot be transitive on $\mathfrak{p}_1$ in this case. On the other hand, if the rank of G is one, then $\overline{\mathfrak{a}_1^+}$ consists of one point, that is, K is transitive on $\mathfrak{p}_1$. We summarize the above discussion as follows:

4.22 Theorem. *Let G be a non-compact semisimple Lie group with finite centre and Cartan decomposition $G = K \exp \mathfrak{p}$.*

(i) *The action of K on $\mathfrak{p}_1$ is transitive if and only if the rank of G/K is 1.*

(ii) *Let Γ be a lattice in G. The geodesic flow on $\Gamma \backslash T^1(G/K)$ is not ergodic if the rank of G/K is ≥ 2.*

This shows that for most locally symmetric Riemannian spaces, that is, for those of rank ≥ 2, the geodesic flow is not ergodic. We shall see that those of rank 1 have an ergodic geodesic flow.

4.23 Example. The above considerations are quite clear in the case $G = \mathrm{SL}(n, \mathbb{R})$ with involution $\theta(A) = (A^{-1})^t$ (see Example 4.5).

The action of $K = \mathrm{SO}(n)$ on $\mathfrak{p}$, the vector space of all symmetric matrices of trace 0, is by conjugation. The Killing form is proportional to the trace form.

The set $\mathfrak{a}$ of diagonal matrices in $\mathfrak{p}$ is a maximal abelian subspace. So, the rank of $\mathrm{SL}(n, \mathbb{R})$ is $n - 1$.

The Weyl group W is the symmetric group on n letters, acting by permutations of the diagonal entries. The roots are λ_{ij}, $1 \leq i \neq j \leq n$,

$$\lambda_{ij}\begin{pmatrix} a_1 & 0 & \cdots & 0 \\ & \cdot & & \\ & & \cdot & \\ 0 & 0 & \cdots & a_n \end{pmatrix} = a_i - a_j.$$

A choice of positive roots is $\Sigma^+ = \{\lambda_{ij} \,|\, i > j\}$. Then

$$\mathfrak{a}^+ = \left\{ \begin{pmatrix} a_1 & 0 & \cdots & 0 \\ & \cdot & & \\ & & \cdot & \\ 0 & 0 & \cdots & a_n \end{pmatrix} \in \mathfrak{a} \,|\, a_1 > a_2 > \cdots > a_n \right\}$$

and

$$\overline{\mathfrak{a}_1^+} = \left\{ \begin{pmatrix} a_1 & 0 & \cdots & 0 \\ & \cdot & & \\ & & \cdot & \\ 0 & 0 & \cdots & a_n \end{pmatrix} \in \mathfrak{a} \,|\, a_1 \geq a_2 \geq \cdots \geq a_n, \quad a_1^2 + \cdots + a_n^2 = 1 \right\}.$$

It is clear that every symmetric matrix (with trace 0) is conjugate under the orthogonal group $SO(n)$ to a unique matrix from $\overline{\mathfrak{a}_1^+}$. The action of $SO(n)$ on $\mathfrak{p}_1$ is transitive if and only if $n = 2$.

4.24 Remark. The Riemannian symmetric spaces G/K for which K is transitive on $T_K(G/K)$ are the two point homogeneous Riemannian symmetric spaces. Recall that a Riemannian manifold M is called *two point homogeneous* if, given points $x, y, x', y' \in M$ with $d(x,y) = d(x',y')$, there exists an isometry g with $gx = x'$ and $gy = y'$ (see [Hel], Chap.X, G).

The Ergodicity in the Rank One Case

We assume that G/K has rank one. So, $\mathfrak{a} = \mathbb{R}H$ for some $H \in \mathfrak{p}$ with $\|H\| = 1$. Recall that $\mathfrak{p}$ is identified with the tangent space to G/K at $o = K$.

4.25 Lemma. *With the above notation, $T^1(G/K) = G \cdot H$. This induces an identification*

$$T^1(G/K) \cong G/M,$$

where M is the centralizer of $\mathfrak{a}$ in K (that is, the set of all $k \in K$ such that $\mathrm{Ad}(k)H = H$). Under this identification, the geodesic flow on

$T^1(G/K)$ *corresponds to the right translations*

$$gM \mapsto g \exp(tH)M, \qquad g \in G, \ t \in \mathbb{R}$$

on G/M.

Proof Since K is transitive on $\mathfrak{p}_1$ (see Theorem 4.22 above), $\mathfrak{p}_1 = \mathrm{Ad}(K)H$.

Let $g \cdot w \in T^1(G/K)$. Choose $k \in K$ such that $w = \mathrm{Ad}(k)H$. Then, for $h = gk$, one has $gK = hK$ and

$$g \cdot w = h \cdot H.$$

This proves the first statement.

Denoting by G_H and K_H be the stabilizers of H in G and K, respectively, one has $G_H = K_H = M$ and the second statement is clear. The third follows from the description of the geodesic flow of G/K (see Theorem 4.14).   □

Observe that, under the above identification, the canonical measure $d\mu$ on $T^1(G/K)$ corresponds to the G–invariant measure on G/M (up to a mutiplicative constant). Indeed, the K-invariant measure dv on $\mathfrak{p}_1 \cong K/M$ corresponds to the K–invariant measure on K/M. As $d\mu$ is locally the product of dv and the G–invariant measure on G/K, the claim is now clear.

Now, let Γ be a lattice in G. Then the geodesic flow on $T^1(\Gamma \backslash G/K)$ corresponds, by the above, to the right translations

$$\Gamma gM \mapsto \Gamma g \exp(tH)M, \qquad g \in G, \ t \in \mathbb{R}$$

on $\Gamma \backslash G/M$. As a consequence, we see that we have to study the unitary represntation of $A = \exp\mathfrak{a}$ defined by the right translations by $\exp tH$ on the Hilbert space $L^2(\Gamma \backslash G/M)$ for $t \in \mathbb{R}$.

We want to apply Mautner's lemma (see Lemma 3.6) to this representation.

4.26 Theorem. *Let* G/K *be a Riemannian symmetric space of rank* 1. *Let* Γ *be a lattice in* G. *The geodesic flow on* $T^1(\Gamma \backslash G/K)$ *is ergodic.*

Proof With the above notation, consider the decomposition of $\mathfrak{g}$ under $\mathrm{ad}\, H$

$$\mathfrak{g} = \mathfrak{g}^0 \oplus \sum_{\lambda \neq 0} \mathfrak{g}^\lambda$$

where $\mathfrak{g}^\lambda$ is the eigenspace to the eigenvalue λ and $\mathfrak{g}^0$ is the centralizer of H in $\mathfrak{g}$. The subspace $\mathfrak{g}^0$ is invariant under the Cartan involution θ. Indeed, more generally, if $X \in \mathfrak{p}$ and $Y \in \mathfrak{g}$ are such that $[X, Y] = \lambda Y$

for some $\lambda \in \mathbb{R}$, then

$$[X, \theta(Y)] = -\theta([X, Y]) = -\lambda Y,$$

so that $\theta \mathfrak{g}^\lambda = \mathfrak{g}^{-\lambda}$. Hence,

$$\mathfrak{g}^0 = \mathfrak{m} \oplus \left(\mathfrak{g}^0 \cap \mathfrak{p}\right),$$

where $\mathfrak{m} = (\operatorname{Ker} \operatorname{ad}(H)) \cap \mathfrak{k}$ is the Lie algebra of M. Now, since $\mathfrak{a} = \mathbb{R}H$ is maximal abelian, $\mathfrak{g}^0 \cap \mathfrak{p} = \mathbb{R}H$ and

$$\mathfrak{g}^0 = \mathfrak{m} \oplus \mathbb{R}H. \tag{3}$$

Fix $\lambda \neq 0$ and $Y \in \mathfrak{g}^\lambda, Y \neq 0$. Let $\mathfrak{g}_{H,Y}$ be the Lie algebra with basis $\{H, Y\}$, and let $G_{H,Y}$ be the corresponding Lie subgroup of G. (Observe that $G_{H,Y}$ is locally isomorphic to the group of triangular matrices in $\mathrm{SL}(2, \mathbb{R})$.) Depending on whether $\lambda > 0$ or $\lambda < 0$, one has

$$\lim_{n \to \pm\infty} \exp(-nH) \exp Y \exp(nH) = e, \tag{4}$$

since

$$\exp(-nH) \exp Y \exp(nH) = \exp(\operatorname{Ad}(\exp(-nH))Y) = \exp(e^{-\lambda n} Y).$$

Denote by π the unitary representation of G on $L^2(\Gamma \backslash G)$ defined by right translation

$$\pi(g)f(\Gamma x) = f(\Gamma x g), \qquad \forall g, x \in G, f \in L^2(\Gamma \backslash G).$$

Since M is compact, $L^2(\Gamma \backslash G/M)$ may be viewed as the subspace of $L^2(\Gamma \backslash G)$ consisting of all $\pi(M)$–invariant functions.

Let $f \in L^2(\Gamma \backslash G/M)$ be invariant under the geodesic flow, that is, f is invariant under $\pi(\exp tH), t \in \mathbb{R}$. Now, due to (4), Mautner's lemma 3.6 applies and shows that f is invariant under $\exp Y$ for any $Y \in \mathfrak{g}^\lambda$ and any $\lambda \neq 0$. By (3), f is also invariant under $\exp Y$ for any $Y \in \mathfrak{g}^0$ (since f is invariant under M). Hence, f is invariant under G and therefore constant. $\qquad\square$

4.27 Example. Let $\mathbf{H}^n$ be the *real hyperbolic space* of dimension n, that is, the set of all points $(x_0, x_1, \ldots, x_n)$ in $\mathbb{R}^{n+1}$ with

$$-x_0^2 + x_1^2 + \cdots + x_n^2 = 1, \quad x_0 > 0.$$

This is the orbit of $a = (1, 0, \ldots, 0)$ under $G = \mathrm{SO}_0(n, 1)$, the connected component of the orthogonal group $O(n, 1)$ of the bilinear form

$$-x_0 y_0 + x_1 y_1 + \cdots + x_n y_n$$

on $\mathbb{R}^{n+1}$. Let J be the $(n+1) \times (n+1)$ diagonal matrix with diagonal elements $-1, 1, \ldots, 1$. Then $O(n, 1)$ is the group of all invertible $(n+1) \times (n+1)$ matrices g with $g^t J g = J$.

The stabilizer of a is $K = SO(n)$, so that $\mathbf{H}^n \cong G/K$. The Cartan involution with fixed point set K is

$$\theta(g) = JgJ, \qquad g \in SO_0(n, 1).$$

The space $\mathfrak{p}$ is the set of all matrices

$$A_{\mathbf{x}} = \begin{pmatrix} 0 & \mathbf{x} \\ \mathbf{x}^t & 0_n \end{pmatrix}, \qquad \mathbf{x} = (x_1, \ldots, x_n) \in \mathbb{R}^n.$$

The Killing form restricted to $\mathfrak{p} \cong \mathbb{R}^n$ is $2n - 2$ times the Euclidean inner product on $\mathbb{R}^n$.

A maximal abelian subspace is

$$\mathfrak{a} = \{A_{\mathbf{x}} | \, \mathbf{x} = (0, \ldots, 0, t), \, t \in \mathbb{R}\}.$$

So, the rank of $\mathbf{H}^n$ is one.

The action of K on $\mathfrak{p}$ is simply

$$\mathrm{Ad}(k)A_{\mathbf{x}} = A_{k \cdot \mathbf{x}}.$$

It is obvious that K acts transitively on the unit sphere of $\mathfrak{p}$.

4.28 Remark. Other examples of Riemannian symmetric spaces of rank one are the complex hyperbolic spaces and the quaternionic hyperbolic spaces, as well as a hyperbolic plane over the Cayley numbers. These examples exhaust the list of the Riemannian (globally) symmetric spaces of negative sectional curvature (see [Hel], Chap. X, Table V).

The hyperbolic space $\mathbf{H}^n$ is a universal covering for every complete Riemannian manifold of constant negative curvature (see [GHL], Chap. III, F). So, we obtain the following corollary.

4.29 Corollary. *The geodesic flow of every complete Riemannian manifold of constant negative curvature and finite volume is ergodic.*

Notes

Concerning Fuchsian groups and fundamental domains, see the classics [Be], [Le] and [Sg2]. A nice introduction to this subject is [Ka]. A comprehensive treatment of symmetric spaces is Helgason's book [Hel]. For more details concerning lattices in Lie groups, we refer to [Rag].

The study of the geodesic flow on a Riemannian manifold has a long history.

It started with Hadamard [Ha] who noticed the instability of geodesics of surfaces of negative curvature. Hedlund [Hel] proved ergodicity of the geodesic flow for compact surfaces with constant negative curvature, by methods of function theory. Introducing a new method, Hopf [Ho1], [Ho2] generalized this result to compact surfaces of variable negative curvature and manifolds of constant negative curvature. (For Hopf's argument, see [KH], 5.4.)

It was the idea of Gelfand–Fomin [GF] to use unitary group representations for the study of the flow in the case of constant curvature. Using the description of the unitary dual of $SL(2, \mathbb{R})$ and $SL(2, \mathbb{C})$, they proved that, for manifolds of dimension 2 or 3 and of constant curvature, the geodesic flow has Lebesgue spectrum. (This means that the unitary representation of $A = \exp \mathfrak{a}$ associated as above to the geodesic flow is a subrepresentation of an infinite multiple of the regular representation of A.)

Following these lines, Mautner [Mau] studied the geodesic flow of locally symmetric Riemannian spaces. His method avoids the (in general, not available) explicit knowledge of the unitary dual of the semisimple Lie groups involved. Our exposition is based on his article [Mau], with simplifications. One may, as done in [Mau], determine in the higher rank case the decomposition of the geodesic flow into ergodic components. Not surprisingly, they correspond to the K-orbits in $\mathfrak{p}_1$. A geometric argument for the transitivity of K on $\mathfrak{p}_1$ in the rank one case is given in [KH], 17.7.

Anosov [An] proved in 1967 the ergodicity of the geodesic flow on any compact Riemannian manifold of negative curvature. The geodesic flow of more general spaces of non-positive curvature – so-called Hadamard spaces – is studied in the monograph [Ba] where a self-contained and complete proof of Anosov's ergodicity theorem is given. For an account on the geodesic flow, see Pansu's notes [Pa].

Chapter III

The Vanishing Theorem of Howe and Moore

Let G be a locally compact group acting on a probability space X, and let π be the associated unitary representation. Recall that the action of G is strongly mixing if every matrix coefficient of π vanishes at infinity on G (see Chap. I, Definition 2.10).

The Vanishing Theorem of Howe and Moore states that matrix coefficients of arbitrary (non trivial) unitary representations of a simple Lie group vanish at infinity. This is a powerful theorem showing that many interesting flows are ergodic (and even strongly mixing). A typical application is as follows. Let G be a simple Lie group. Let Γ be a lattice and let H be a non-compact closed subgroup of G. Then the action of H on $\Gamma \backslash G$ by right translations is ergodic. As we saw in the previous chapter, the geodesic flow of a locally symmetric space is related to flows of this type.

A more general application is Moore's ergodicity theorem which gives optimal conditions under which the action of a subgroup of a semisimple Lie group is ergodic or even strongly mixing (see Theorems 2.1, 2.5).

The theorem of Howe and Moore is first proved for $SL(2, \mathbb{R})$. The proof we present, due to R. Howe, is remarkable in that it essentially uses no representation theory. Again, a crucial rôle is played by Mautner's lemma (Chap. II, 3.6)

The extension to general semisimple Lie groups is a technical matter: one has to use sufficiently many copies of $SL(2, \mathbb{R})$ inside the given group. Here we follow [Ve1].

A nice application of strong mixing appeared in [EM]. It is a formula for the asymptotic behaviour of the number of lattice points in a ball in the hyperbolic plane (Theorem 3.5).

As another application, we discuss a natural strengthening of strong mixing called mixing of all orders. It is an old (and still open) problem in ergodic theory whether a transformation (that is, an action of the integers $\mathbb{Z}$) which is strongly mixing is necessarily mixing of all orders. We treat the counterexample, discovered by F. Ledrappier [Ld], for an action of $\mathbb{Z}^2$ (Proposition 4.4 and Corollary 4.10).

Following S. Mozes [Mz1], we show that ergodic actions of semisimple Lie groups are automatically mixing of all orders (Theorem 4.11).

§1 Howe–Moore's Theorem

Let G be a locally compact group, and let $(\pi, \mathcal{H}_\pi)$ be a strongly continuous unitary representation of G. The functions

$$\varphi_{\xi\,\eta} : G \to \mathbb{C}, \quad g \mapsto \langle \pi(g)\xi, \eta \rangle,$$

where ξ, η are vectors in $\mathcal{H}_\pi$, are the *matrix coefficients* associated to π.

Let G be a connected semisimple Lie group with Lie algebra $\mathfrak{g}$. By standard structure theory, $\mathfrak{g}$ is the direct sum

$$\mathfrak{g} = \mathfrak{g}_1 \oplus \cdots \oplus \mathfrak{g}_n$$

of its simple ideals $\mathfrak{g}_1, \ldots, \mathfrak{g}_n$, the corresponding connected simple normal subgroups $S_1, \ldots, S_n$ of G are closed and G is the (not necessarily direct) product

$$G = S_1 \cdots S_n.$$

Observe that every element from S_i commutes with every element from S_j for all $i \neq j$ and, hence, $S_i \cap \prod_{j \neq i} S_j$ lies in the discrete centre of G. We shall refer to $S_1, \ldots, S_n$ as the *simple factors* of G. We say that G *has no compact factor* if no S_i is compact.

We formulate Howe–Moore's theorem. Recall that a continuous function $f : X \to \mathbb{C}$ on a locally compact space X vanishes at infinity if $\lim_n f(x_n) = 0$, for every sequence $\{x_n\}_n$ in X diverging to ∞ (that is, with no limit point in X).

1.1 Theorem (Howe–Moore Vanishing Theorem). *Let G a connected semisimple Lie group with finite centre. Let $(\pi, \mathcal{H}_\pi)$ be a strongly continuous unitary representation of G. Assume that the restriction of π to any non-compact simple factor S_i of G has no non-trivial invariant vector. Then all the matrix coefficients of π vanish at infinity.*

The assumption made on π is clearly necessary. Indeed, if $\xi \neq 0$ is fixed by a non-compact closed subgroup H then, with the above notation, the matrix coefficient $\varphi_{\xi\,\xi}$ is constant on H and, hence, cannot vanish at infinity.

The proof of the theorem is in several steps. We first need a refinement of the Cartan decomposition from Chapter II.

The Cartan Decomposition Revisited

Let G be a semisimple non-compact Lie group, with Lie algebra $\mathfrak{g}$. As mentioned in the previous chapter (see the remarks before Chap. II, Example 4.11), there exists a Cartan involution θ on $\mathfrak{g}$. Let $\mathfrak{k}$ and $\mathfrak{p}$ be the eigenspaces to the eigenvalues 1 and -1, respectively. Then K, the subgroup corresponding to $\mathfrak{k}$, is a maximal compact subgroup, and one has the Cartan decomposition

$$G = K \exp \mathfrak{p}$$

of G.

Fix a maximal abelian subalgebra $\mathfrak{a}$ of $\mathfrak{p}$. (Observe that, since G is not compact, $\mathfrak{p}$ and $\mathfrak{a}$ are not trivial.) Let A be the subgroup corresponding to $\mathfrak{a}$.

Let Σ be the set of roots of $\mathfrak{a}$, and let Σ^+ be the subset of positive roots (with respect to some ordering of $\mathfrak{a}^*$). Let

$$A^+ := \{\exp H \mid H \in \mathfrak{a}^+\},$$

where $\mathfrak{a}^+ = \{H \mid \lambda(H) > 0, \ \forall \lambda \in \Sigma^+\}$ is the positive Weyl chamber. One has the following decomposition, also called the Cartan decomposition of G,

$$G = K \cdot \overline{A^+} \cdot K,$$

where $\overline{A^+}$ is the closure of A^+ (see [Hel], Chap. IX, Theorem 1.1). Observe that this decomposition follows from the previous Cartan decomposition $G = K \exp \mathfrak{p}$ and from the fact that $\mathfrak{p} = \bigcup_{k \in K} \mathrm{Ad}(k)\mathfrak{a}$ (see remarks following Chap. II, Remark 4.20)

1.2 Example. In the case $G = \mathrm{SL}(n, \mathbb{R})$ (with the choices made in Chap. II, Example 4.5), K is $\mathrm{SO}(n)$, A is the subgroup of diagonal matrices with positive eigenvalues and $\overline{A^+}$ is the subset of A of all diagonal matrices with eigenvalues $a_1 \geq a_2 \geq \cdots \geq a_n$.

The following elementary fact is a consequence of the above Cartan decomposition.

1.3 Lemma. *Let $(\pi, \mathcal{H}_\pi)$ be a strongly continuous unitary representation of G. Suppose that, for all matrix coefficients $\varphi_{\xi\,\eta}$ of π and for all sequences $\{a_n\}_n$ in $\overline{A^+}$ with $\lim_{n \to \infty} a_n = \infty$, one has*

$$\lim_{n \to \infty} \varphi_{\xi\,\eta}(a_n) = 0.$$

Then all matrix coefficients of π vanish at infinity on G.

Proof Assume, by contradiction, that there exists a sequence $\{g_n\}_n$ in G diverging to infinity such that

$$\lim_{n \to \infty} \varphi_{\xi\,\eta}(g_n) \neq 0.$$

Let $g_n = k_n a_n h_n$, where $k_n, h_n \in K$, $a_n \in \overline{A^+}$, be a Cartan decomposition of g_n. Clearly, $\lim_{n \to \infty} a_n = \infty$.

As K is compact, upon passing to a subsequence, we may assume that

$\pi(h_n)\xi$ and $\pi(k_n^{-1})\eta$ converge (in norm) to vectors $\bar{\xi}$ and $\bar{\eta}$ in $\mathcal{H}_\pi$. Then

$$\varphi_{\xi\,\eta}(k_n a_n h_n) = \left(\varphi_{\xi\,\pi(k_n^{-1})\eta}(a_n h_n) - \varphi_{\xi\,\bar{\eta}}(a_n h_n)\right)$$
$$+ \left(\varphi_{\pi(h_n)\xi\,\bar{\eta}}(a_n) - \varphi_{\bar{\xi}\,\bar{\eta}}(a_n)\right) + \varphi_{\bar{\xi}\,\bar{\eta}}(a_n)$$

The first two summands tend to zero as $g_n \to \infty$. Hence,

$$\lim_{n\to\infty} \varphi_{\bar{\xi}\,\bar{\eta}}(a_n) \neq 0.$$

This is a contradiction. □

The Mautner Lemma Revisited

Let $(\pi, \mathcal{H}_\pi)$ be a strongly continuous unitary representation of an arbitrary locally compact group G. Let $\xi \in \mathcal{H}_\pi$, and let $\{g_n\}$ be a sequence in G. The sequence $\pi(g_n)\xi$ is norm bounded in $\mathcal{H}_\pi$. Hence, $\{\pi(g_n)\xi\}_{n\in\mathbb{N}}$ has accumulation points in $\mathcal{H}_\pi$ with respect to the *weak* topology. If one could prove that 0 is the only accumulation point, then $\lim \varphi_{\xi\,\eta}(g_n) = 0$. The idea is to find large subgroups leaving ξ_0 invariant.

For a sequence $\alpha = \{a_n\}_n$ in G, let

$$U_\alpha^+ := \{g \in G|\ e \text{ is an accumulation point of } \{a_n^{-1}g a_n\}_{n\in\mathbb{N}}\},$$

and let N_α^+ be the closed subgroup generated by U_α^+. The following is a stronger version of the Mautner Lemma given in Chap. II, Lemma 3.6, with essentially the same proof.

1.4 Theorem (Mautner Phenomenon). *Let G be a locally compact group and $(\pi, \mathcal{H}_\pi)$ a strongly continuous unitary representation of G. Let $\alpha = \{a_n\}_n$ be a sequence in G and let $\xi, \xi_0 \in \mathcal{H}_\pi$ be such that*

$$\lim_{n\to\infty} \pi(a_n)\xi = \xi_0.$$

Then

$$\pi(x)\xi_0 = \xi_0$$

for all $x \in N_\alpha^+$.

Proof Let $\{a_{n_k}\}_{k\in\mathbb{N}}$ be a subsequence of α such that $a_{n_k}^{-1}x a_{n_k}$ converges to e. Then

$$|\langle \pi(x)\xi_0, \eta\rangle - \langle \xi_0, \eta\rangle| = \lim_{k\to\infty} |\langle \pi(x a_{n_k})\xi, \eta\rangle - \langle \pi(a_{n_k})\xi, \eta\rangle|$$
$$= \lim_{k\to\infty} |\langle \pi(a_{n_k}^{-1}x a_{n_k})\xi, \pi(a_{n_k})^{-1}\eta\rangle - \langle \xi, \pi(a_{n_k}^{-1})\eta\rangle|$$
$$\leq \lim_{k\to\infty} \|\left(\pi(a_{n_k}^{-1}x a_{n_k}) - \mathrm{id}_{\mathcal{H}_\pi}\right)\xi\|\|\eta\| = 0,$$

by unitarity of π, the Cauchy–Schwarz inequality and strong continuity of π. $\square$

1.5 Example. Let $G = \mathrm{SL}(2, \mathbb{R})$, with

$$A^+ = \left\{ \begin{pmatrix} \alpha & 0 \\ 0 & \frac{1}{\alpha} \end{pmatrix} \mid \alpha > 1 \right\},$$

$$\overline{A^+} = A^+ \cup \{e\}$$

and

$$N = \left\{ \begin{pmatrix} 1 & x \\ 0 & 1 \end{pmatrix} \mid x \in \mathbb{R} \right\}.$$

A sequence $\alpha = \left\{ \begin{pmatrix} \alpha_n & 0 \\ 0 & \alpha_n^{-1} \end{pmatrix} \right\}_{n \in \mathbb{N}}$ in $\overline{A^+}$ converges to infinity if and only if

$$\lim_{n \to \infty} \alpha_n = \infty.$$

It is straightforward to verify that $N_\alpha^+ = N$.

Proof of the Vanishing Theorem for Groups Locally Isomorphic to $\mathrm{SL}(2, \mathbb{R})$

Let G be a group locally isomorphic to $\mathrm{SL}(2, \mathbb{R})$, with finite centre. The Lie algebra of G is $\mathfrak{g} = \mathfrak{sl}(2, \mathbb{R})$, with standard basis

$$H = \begin{pmatrix} 1 & 0 \\ 0 & -1 \end{pmatrix}, \quad X = \begin{pmatrix} 0 & 1 \\ 0 & 0 \end{pmatrix}, \quad Y = \begin{pmatrix} 0 & 0 \\ 1 & 0 \end{pmatrix}.$$

As above, $A = \exp_G \mathbb{R}H$, and $A^+ = \exp_G \mathbb{R}^+ H$. Set

$$N = \exp_G \mathbb{R}X \quad \text{and} \quad N^- = \exp_G \mathbb{R}Y.$$

Let $(\pi, \mathcal{H}_\pi)$ be a unitary representation of G without non-zero invariant vectors. Fix a vector ξ in $\mathcal{H}_\pi$, and let $\alpha = \{a_n\}_n$ be a sequence in $\overline{A^+}$, diverging to infinity. Then N is contained in N_α^+ . Indeed, if $a_n = \exp_G(\alpha_n H)$, then $\lim_n \alpha_n = +\infty$, and

$$\lim_n \exp_G(-\alpha_n H) \exp_G tX \exp(\alpha_n H) = \lim_n \exp_G(e^{-2\alpha_n} tX) = e,$$

as $\mathrm{ad}\,(H)X = 2X$.

Now, let $\xi_0 \in \mathcal{H}_\pi$ be an accumulation point of $\{\pi(a_n)\}_n$. Then ξ_0 is invariant under N, by the Mautner Phenomenon (Theorem 1.4). The following lemma shows that then ξ_0 is invariant under the whole group G. Hence, $\xi_0 = 0$. As discussed above, this together with Lemma 1.3 implies that $\varphi_{\xi \eta}$ vanishes at infinity for any vector η and concludes the proof.

1.6 Theorem. *Let G be a group locally isomorphic to $\mathrm{SL}(2,\mathbb{R})$ and with finite centre, and let $(\pi, \mathcal{H}_\pi)$ be a unitary representation of G. Let ξ_0 be a vector in $\mathcal{H}_\pi$ which is invariant under $N = \exp_G \mathbb{R} X$. Then ξ_0 is invariant under G.*

Proof (i) We first deal with the case $G = \mathrm{SL}(2,\mathbb{R})$. We claim that ξ_0 is invariant under A. By Mautner's lemma, this will imply that ξ_0 is invariant under N^-. Since $\mathrm{SL}(2,\mathbb{R})$ is generated by N and N^-, this will finish the proof.

Consider the continuous function

$$\varphi(g) = \langle \pi(g)\xi_0, \xi_0 \rangle, \qquad g \in \mathrm{SL}(2,\mathbb{R})$$

(the positive definite function associated to ξ_0). Then φ is N-bi-invariant, that is, φ is constant on every double coset NgN, $g \in \mathrm{SL}(2,\mathbb{R})$.

Fix any sequence $\lambda_n \in \mathbb{R}, \lambda_n \neq 0$, with $\lambda_n \to 0$, and define

$$g_n = \begin{pmatrix} 0 & -\lambda_n^{-1} \\ \lambda_n & 0 \end{pmatrix}.$$

Let $a = \begin{pmatrix} \alpha & 0 \\ 0 & \alpha^{-1} \end{pmatrix} \in A$. Then,

$$\begin{pmatrix} 1 & \alpha\lambda_n^{-1} \\ 0 & 1 \end{pmatrix} g_n \begin{pmatrix} 1 & \alpha^{-1}\lambda_n^{-1} \\ 0 & 1 \end{pmatrix} = \begin{pmatrix} \alpha & 0 \\ \lambda_n & \alpha^{-1} \end{pmatrix}.$$

Hence, $\varphi(a) = \lim_n \varphi(g_n)$ for all $a \in A$. Thus, φ is constant on A, that is, for all $a \in A$,

$$\langle \pi(a)\xi_0, \xi_0 \rangle = \|\xi_0\|^2. \tag{$*$}$$

Hence, by the Cauchy–Schwarz inequality (equality case), $\pi(a)\xi_0$ is a scalar multiple $\chi(a)\xi_0$ of ξ_0. Equation $(*)$ shows that, necessarily, $\pi(a)\xi_0 = \xi_0$, for all $a \in A$. This finishes the proof in the case $G = \mathrm{SL}(2,\mathbb{R})$.

(ii) The case $G = \mathrm{PSL}(2,\mathbb{R}) = \mathrm{SL}(2,\mathbb{R})/\{\pm I\}$ follows immediately. Indeed, π may be lifted to a representation of $\mathrm{SL}(2,\mathbb{R})$.

(iii) Let $p : G \to \mathrm{PSL}(2,\mathbb{R})$ be an arbitrary finite covering. Then $Z = \mathrm{Ker}\, p$ is the centre of G and $G/Z \cong \mathrm{PSL}(2,\mathbb{R})$. Let

$$\mathcal{H}_\pi = \mathcal{H}_1 \oplus \cdots \oplus \mathcal{H}_r$$

be the decomposition of $\mathcal{H}_\pi$ into Z-isotypic components. Since Z is central, each $\mathcal{H}_i$ is invariant under G. Let $\xi_0^{(1)}, \ldots, \xi_0^{(r)}$ be the components of ξ_0 with respect to this decomposition. Observe that each $\xi_0^{(i)}$ is fixed under N.

For each i,

$$g \mapsto \psi_i(g) := |\langle \pi(g)\xi_0^{(i)}, \xi_0^{(i)} \rangle|^2$$

is a positive definite function on G (a coefficient of the tensor product $\pi \otimes \overline{\pi}$). Clearly, ψ_i is constant on every Z-coset and may therefore be viewed as a positive definite function on $\mathrm{PSL}(2,\mathbb{R})$. As ψ_i is constant on N, it follows from (ii) above that ψ_i is constant on G. The Cauchy–Schwarz inequality shows (as above) that, for all $g \in G$,

$$\pi(g)\xi_0^{(i)} = \chi(g)\xi_0^{(i)},$$

for some $\chi(g) \in \mathbb{C}$ with $|\chi(g)| = 1$. But then χ has to be a unitary character of G. The only such character is 1. Indeed, the kernel H of χ is a normal subgroup of G. Hence, if H were proper, then H would be contained in Z and G/Z would be abelian. This is not the case. Hence, $\chi = 1$, that is, ξ_0 is fixed under G. $\square$

1.7 Remarks. (i) A more geometric – and slightly longer – proof of the above theorem for $G = \mathrm{SL}(2, \mathbb{R})$ may be found in [Zi], proof of 2.4.2. Another proof based on consideration of the derived representation of the Lie algebra is given in [BeM], Theorem 2.5.

(ii) The proof of the above theorem is also valid, without any change, for $G = \mathrm{SL}(2, \mathbb{F})$ over a non-discrete topological field $\mathbb{F}$.

(iii) The theorem is true even if G has infinite centre, with the same arguments. One only has to use – instead of a finite sum – a direct integral decomposition of $\mathcal{H}_\pi$ under the action of the centre Z.

(iv) The theorem immediately implies the ergodicity of the horocyclic flow on surfaces of constant negative curvature (see Chap. IV, Theorem 1.3).

Proof of the Vanishing Theorem for General Semisimple Groups

Let G be an arbitrary semisimple Lie group with finite centre. For $b \in G$ and $\alpha = \{b^n\}_n$, we shall write U_b^+ and N_b^+ instead of U_α^+ and N_α^+. (Recall that, for a sequence $\alpha = \{a_n\}_n$, U_α^+ is the set of all $g \in G$ such that e is an accumulation point of $\{a_n^{-1} g a_n\}_n$ and that N_α^+ is the subgroup generated by U_α^+.)

1.8 Lemma.

(i) *For $b \in A \setminus \{e\}$, let G_b be the closed subgroup generated by N_b^+ and $N_{b^{-1}}^+$. Then G_b is a non-discrete normal subgroup of G.*

(ii) *For every sequence α in $\overline{A^+}$ converging to infinity, there exists $b \in \overline{A^+} \setminus \{e\}$ with $N_\alpha^+ = N_b^+$.*

Proof (i) Let

$$\mathfrak{g} = \sum_{\lambda \in \Sigma} \mathfrak{g}^\lambda \ \oplus \ \mathfrak{g}^0$$

be the root space decomposition of $\mathfrak{g}$ under $\mathfrak{a}$. Let $H \in \mathfrak{a}$ be such that $b = \exp_G H$. Define

$$\mathfrak{g}_b^- := \sum_{\lambda(H)<0} \mathfrak{g}^\lambda, \quad \mathfrak{g}_b^0 := \sum_{\lambda(H)=0} \mathfrak{g}^\lambda, \quad \mathfrak{g}_b^+ := \sum_{\lambda(H)>0} \mathfrak{g}^\lambda.$$

Then $\mathfrak{g} = \mathfrak{g}_b^- \oplus \mathfrak{g}_b^0 \oplus \mathfrak{g}_b^+$. Let $\mathfrak{g}_b$ be the subalgebra of $\mathfrak{g}$ generated by $\mathfrak{g}_b^+$ and $\mathfrak{g}_b^-$. Since

$$[\mathfrak{g}^\lambda, \mathfrak{g}^\mu] \subseteq \mathfrak{g}^{\lambda+\mu}, \quad \forall \lambda, \mu \in \Sigma,$$

one has $[\mathfrak{g}_b^0, \mathfrak{g}_b] \subseteq \mathfrak{g}_b$. Hence, $\mathfrak{g}_b$ is an ideal in $\mathfrak{g}$.

On the other hand, since $b^n \exp_G X b^{-n} = \exp_G(\mathrm{Ad}(b)^n X)$, $\exp_G \mathfrak{g}_b^+$ is contained in N_b^+ and $\exp_G \mathfrak{g}_b^-$ is contained in N_{b-1}^+. Hence,

$$\exp_G(\mathfrak{g}_b) \subseteq G_b. \tag{$**$}$$

Since the exponential mapping is a local diffeomorphism, every neighbourhood of e contains a neighbourhood of the form

$$U = U^- \cdot U^0 \cdot U^+,$$

where $U^\pm$ and U^0 are neighbourhoods of e in $\exp_G \mathfrak{g}_b^\pm$ and in $\exp_G \mathfrak{g}^0$, respectively.

Now let $x \in U_b^+$. For some n large enough, we have

$$b^{-n} x b^n \in U_b^+ \cap U = U^+.$$

This implies that $x \in \exp_G(\mathfrak{g}_b^+)$. Thus, $U_b^+ \subseteq \exp_G(\mathfrak{g}_b^+)$. Similarly, $U_{b-1}^+ \subseteq \exp_G(\mathfrak{g}_b^-)$. This and $(**)$ show that G_b is generated by $\exp_G(\mathfrak{g}_b)$.

Observe that G_b is non-discrete, that is, $\mathfrak{g}_b \neq 0$. Indeed, otherwise H would be in the centre of $\mathfrak{g}$. Hence, $H = 0$ and $b = e$.

(ii) Let $\alpha = \{\exp_G(a_n)\}_n$ with $a_n \in \overline{\mathfrak{a}^+}$ and $\lim_{n\to\infty} \exp_G(a_n) = \infty$. Choose a basis $R \subseteq \Sigma^+$ of the root system Σ of $\mathfrak{a}$. (Any root is a linear combination of roots from R with either all coefficients being non-negative integers or all coefficients being non-positive integers.) There exists some $\lambda \in R$ with

$$\limsup_{n\to\infty} \lambda(a_n) = \infty. \tag{$***$}$$

Indeed, otherwise $\mathrm{Ad}(\exp(a_n))$ would be bounded in $\mathrm{Ad}(G)$. Since $\mathrm{Ad}(G)$ is a quotient of G by a finite subgroup, this would be a contradiction to $\lim_{n\to\infty} \exp_G(a_n) = \infty$.

Let R_α be the set of all $\lambda \in R$ such that $(***)$ holds. Now let $H \in \overline{\mathfrak{a}^+}$ be defined by

$$\lambda(H) = \begin{cases} 1 & \text{if } \lambda \in R_\alpha \\ 0 & \text{if } \lambda \in R \setminus R_\alpha \end{cases}$$

and let $b := \exp_G H$.

Clearly, $b \in \overline{A^+}$ and $b \neq e$. Observe that, for $\lambda \in \Sigma$, one has $\lambda(H) > 0$ if and only if λ is a positive root not contained in the vector space spanned by $R \setminus R_\alpha$.

Using the notations of the proof of (i), $\exp_G(\mathfrak{g}_b^+) \subseteq U_\alpha^+$. This follows from the fact that $\exp_G \mathfrak{g}^\lambda \subseteq U_\alpha^+$ if $\lambda(H) > 0$.

We claim that also $U_\alpha^+ \subseteq \exp_G(\mathfrak{g}_b^+)$. As N_b^+ is the subgroup generated by $\exp_G(\mathfrak{g}_b^+)$ (see (i) above), this will complete the proof. To show this, let $x \in U_\alpha^+$. Then, with the above neighbourhood $U = U^- \cdot U^0 \cdot U^+$, for n large enough,

$$\exp_G(a_n)^{-1} x \exp_G(a_n) = \exp X_n^+ \exp X_n^0 \exp X_n^-$$

for $X_n^\pm \in \mathfrak{g}_b^\pm$ and $X_n^0 \in \mathfrak{g}_b^0$ with $\lim_n X_n^\pm = \lim_n X_n^0 = 0$. Hence,

$$x = \exp_G(\mathrm{Ad}(a_n) X_n^+) \exp_G(\mathrm{Ad}(a_n) X_n^0) \exp_G(\mathrm{Ad}(a_n) X_n^-).$$

Now, $\lim_n \mathrm{Ad}(a_n) X_n^0 = \lim_n \mathrm{Ad}(a_n) X_n^- = 0$, since $\mathrm{Ad}(a_n)$ is bounded on $\mathfrak{g}_b^0$ and $\mathfrak{g}_b^-$. Thus,

$$x = \lim_n \exp_G(\mathrm{Ad}(a_n) X_n^+)$$

and $x \in \exp_G(\mathfrak{g}_b^+)$.   □

Now we are ready to give the proof of the Vanishing Theorem.

Proof of Theorem 1.1. Let π be a unitary representation of G. Assume that the restriction of π to any non-compact simple factor has no non-trivial invariant vector.

First of all, we may assume that G has no compact factor. Indeed, if the restriction of every matrix coefficient of π to the product of the non-compact simple factors of G vanishes at infinity, then every matrix coefficient of π vanishes at infinity on G. Let $\varphi_{\xi\eta}$ be a matrix coefficient of π. Let $\alpha = \{a_n\}_n$ be a sequence in $\overline{A^+}$ which tends to infinity, and let ξ_0 be a vector in $\mathcal{H}_\pi$ such that

$$\lim_{n \to \infty} \pi(a_n)\xi = \xi_0.$$

We claim that ξ_0 is invariant under some simple factor of G. This will imply that $\xi_0 = 0$ and will finish the proof.

By Mautner's lemma, ξ_0 is invariant under N_α^+. The previous lemma shows that $N_\alpha^+ = N_b^+$ for a suitable $b \in \overline{A^+} \setminus \{e\}$.

Fix $\lambda \in \Sigma^+$ with $\mathfrak{g}^\lambda \subseteq \mathfrak{g}_b^+$, and let $X \in \mathfrak{g}^\lambda \setminus \{0\}$ (notation being as in the proof of the previous lemma). Let $Q_\lambda \in \mathfrak{a}$ be defined by

$$\lambda(H) = \kappa(H, Q_\lambda), \qquad \forall H \in \mathfrak{a}$$

(κ denotes the Killing form). The Lie algebra $\mathfrak{g}_X$ generated by X, $\theta(X)$ and Q_λ is isomorphic to $\mathfrak{sl}(2, \mathbb{R})$. Indeed,

$$[X, \theta(X)] \in \mathfrak{g}^0 \cap \mathfrak{p} = \mathfrak{a},$$

and, for all $H \in \mathfrak{a}$,

$$\kappa(H, [X, \theta(X)]) = \kappa([H, X], \theta(X)) = \lambda(H)\kappa(X, \theta(X)).$$

Hence, as $\kappa(X, \theta(X)) \neq 0$,

$$[X, \theta(X)] = \frac{1}{\kappa(X, \theta(X))} Q_\lambda.$$

This shows that $\mathfrak{g}_X$ is isomorphic to $\mathfrak{sl}(2, \mathbb{R})$.

Let G_X be the corresponding subgroup of G. Since G has finite centre, the kernel of the adjoint representation Ad of G has finite kernel. The group $\mathrm{Ad}(G_X)$ is a simple linear Lie group. As such, it has finite centre and is closed in $\mathrm{GL}(\mathfrak{g})$ (see [Hel], Chap III. Exercise B7). This clearly implies that G_X has finite centre and is closed in G.

Now, ξ_0 is invariant under $N = \exp_G \mathbb{R}X$. Hence, by Theorem 1.6, ξ_0 is invariant under G_X. Since this holds for any X and λ, we conclude that ξ_0 is invariant under G_b. (Observe that $\theta(\mathfrak{g}_b^+) = \mathfrak{g}_b^- = \mathfrak{g}_{b-1}^+$.) By the previous lemma, G_b is a non-discrete normal subgroup of G. Hence, G_b contains some simple factor of G. This finishes the proof. $\qquad\square$

§2 Moore's Ergodicity Theorems

Moore's ergodicity theorems give very general conditions under which a subgroup of a semisimple group acts ergodically. They were originally motivated by the study of the geodesic flow on symmetric spaces initiated by Gelfand and Fomin [GF] (see II.3 and II.4). Howerever, in contrast to [GF], the explicit knowledge of the dual of the acting semisimple group is not needed in Moore's approach [Mo1], [Mo2]. We will deduce the theorems as corollaries of the Howe–Moore vanishing theorem (see also [Zi]).

As in the previous section, let G be a connected semisimple Lie group with finite centre. Then $G = S_1 \cdots S_n$, where $S_1, \ldots, S_n$ are the simple factors of G.

Here is the "concrete" version of Moore's Ergocity Theorem.

2.1 Theorem (Moore's Ergodicity Theorem). *Let G be a semisimple Lie group with finite centre. Let G act on a probability space (X, μ), and assume that the restriction of this action to any simple non-compact factor of G is ergodic. Let H be a subgroup of G with a non-compact closure. Then the action of H on X is strongly mixing (and, hence, ergodic).*

Proof Let $(\pi, L_0^2(X, \mu))$ be the associated representation of G. The restriction of π to any simple factor non-compact S_i has no non-trivial invariant vector, since S_i acts ergodically on X. Therefore, by Howe–Moore's theorem, all matrix coefficients of π vanish at infinity. Assume that there is a non-trivial $\xi_0 \in L^2(X, \mu)$, stabilized by H. Then ξ_0 is stabilized by $\overline{H}$ and the matrix coefficient $\varphi_{\xi_0\,\xi_0}$ is constant on the non-compact closed set $\overline{H}$, a contradiction.   □

2.2 Corollary. *If a simple Lie group G with finite centre acts ergodically on a probability space X, then every subgroup of G with a non-compact closure is strongly mixing on X.*

Using the description of the geodesic flow a Riemannian symmetric space from Chap. II, Lemma 4.25, and the remark before Chap. II, Corollary 4.29, we obtain the following:

2.3 Corollary. *Let G/K be a Riemannian symmetric space of rank 1. Let Γ be a lattice in G. The geodesic flow of $\Gamma \backslash G/K$ is strongly mixing. In particular, this is true for the geodesic flow of any Riemannian manifold of constant negative curvature and finite volume.*

We now give a more general, "abstract" version of Moore's Ergodicity Theorem.

2.4 Definition. Let G be a connected semisimple Lie group, with simple factors $S_1, \ldots, S_n$ and centre Z. Let p be the canonical projection of G onto the adjoint group G/Z. Then G/Z is the direct product

$$G/Z = p(S_1) \times \cdots \times p(S_n).$$

Let p_i denote the projection of G onto $p(S_i)$. A subgroup H of G is called *totally non-compact* if $H_i := p_i(H)$ has non-compact closure for all $1 \leq i \leq n$. Observe that this condition is never satisfied if G has a compact factor.

2.5 Theorem (Moore's Ergodicity Theorem). *Let G be a connected semisimple Lie group with finite centre and without compact factors. Let H be a totally non-compact subgroup of G, and let $(\pi, \mathcal{H}_\pi)$ be a unitary representation of G.*

(i) *If $\xi \in \mathcal{H}_\pi$ is invariant under H, then ξ is invariant under G. More generally, if V is an H-invariant finite-dimensional subspace of $\mathcal{H}_\pi$, then G acts trivially on V.*

(ii) *If H is closed, all matrix coefficients of the representation $(\pi|_H, \mathcal{H}_\pi)$ vanish at infinity, provided that the restriction of π to each simple factor does not contain the trivial representation.*

Proof Upon passing to a finite covering group of G, we may assume that G is the *direct* product of its simple factors

$$G = S_1 \times \cdots \times S_n.$$

For each subset I of $\{1,\ldots,n\}$, define $(\pi_I,\mathcal{H}_{\pi_I})$ to be the subrepresentation of π whose kernel contains $\prod_{i\in I}S_i$, but no S_j for $j\notin I$. So, $\mathcal{H}_{\pi_I}$ is the subspace of vectors which are fixed by all S_i, $i\in I$, but not by any of the others. Define $\pi_I=0$, if this space is $\{0\}$. Then, clearly, we have an orthogonal decomposition

$$\pi=\bigoplus_{I\subseteq\{1,\ldots,n\}}\pi_I.$$

By Howe–Moore's theorem, all matrix coefficients of π_I, considered as a representation of $\prod_{j\notin I}S_j$, vanish at infinity for $I\neq\{1,\ldots,n\}$. Let H_I be the closure of the image of H in $\prod_{j\notin I}S_j$. Since H is totally non-compact, H_I is non-compact. Thus, all matrix coefficients of $\pi_I|_{H_I}$ are in $C_0(H_I)$ for $I\neq\{1,\ldots,n\}$.

(i) Let V be a finite-dimensional H-invariant subspace of $\mathcal{H}_\pi$, and let V_I be the orthogonal projection of V on $\mathcal{H}_{\pi_I}$. Then V_I is an H_I-invariant finite-dimensional subspace and defines a subrepresentation of $\pi_I|_{H_I}$. Now, matrix coefficients associated to finite-dimensional representations of a non-compact group cannot vanish at infinity (see Chap. I, Remark 2.15 (iii)). Hence, the above implies that $V_I=\{0\}$ for all $I\neq\{1,\ldots,n\}$. Therefore, $V=V_{\{1,\ldots,n\}}$ is pointwise fixed under G.

(ii) This follows immediately from Howe–Moore's theorem. $\qquad\square$

2.6 Remark. Non-compactness of H is not sufficient in (i) of the above theorem. Let, for instance, $G=\mathrm{SL}(2,\mathbb{R})\times\mathrm{SL}(2,\mathbb{R})$, $H=\mathrm{SO}(2)\times\mathrm{SL}(2,\mathbb{R})$, and choose $\pi=\sigma\otimes 1$ for a non-trivial representation σ of $\mathrm{SL}(2,\mathbb{R})$ with non-trivial $\mathrm{SO}(2)$-fixed vectors. Then, the restriction of π to H is trivial.

We now give as an application some ergodicity results about actions on homogeneous spaces without invariant measures.

Let H be a closed subgroup of a locally compact group G. The homogeneous space $X=G/H$ always has a quasi-invariant measure μ. The measure class of μ is unique. For all this, see [Re], Chap. 8 and [Bo2], Chap. VII. We shall always choose μ so that Weil's formula holds (see remarks before Chap. II, Definition 2.1). The following duality result has been first observed by C.C. Moore [Mo1].

2.7 Lemma (Moore Duality Theorem). *Let H_1,H_2 be closed subgroups of a locally compact group G. Then H_1 is ergodic on G/H_2 if and only if H_2 is ergodic on G/H_1.*

Proof Assume that H_1 is not ergodic on G/H_2. Then there exists a bounded measurable function f on G/H_2 which is not constant almost everywhere and such that $f(h_1\cdot x)=f(x)$ for all $h_1\in H_1$ and $x\in G/H_2$.

Let F be the measurable function on G defined by

$$F(g) = f(gH_2), \qquad \forall g \in G.$$

Then $F(h_1 g h_2) = F(g)$ for all $g \in G$, $h_1 \in H_1$ and $h_2 \in H_2$. By the lemma below, F is not constant almost everywhere on G. Clearly, F factorizes to a bounded measurable function f' on $H_1 \setminus G$ defined by $f'(H_1 g) = F(g)$ for all $g \in G$. Again by the lemma below, f' is not constant almost everywhere. As $f'(xh_2) = f'(x)$ for all $x \in H_1 \setminus G$ and $h_2 \in H_2$, it follows from Chap. I, Theorem 1.3, that H_2 is not ergodic on H_1/G. Now, the mapping

$$G/H_1 \to H_1 \setminus G, \qquad gH_1 \mapsto H_1 g^{-1}$$

intertwines the natural actions of G on G/H_1 and on $H_1 \setminus G$ and sends sets of measure zero to sets of measure zero. Hence, H_2 is not ergodic on G/H_1. $\square$

2.8 Lemma. *Let H be a closed subgroup of a locally compact group G, and let f be a bounded measurable function on G/H. Let F be the measurable function on G defined by $F(g) = f(gH)$ for all $g \in G$. Then $f = 0$ almost everywhere on G/H if and only if $F = 0$ almost everywhere on G.*

Proof Let $\varphi \in C_c(G)$. Then, $F\varphi$ is integrable on G and, by Weil's formula (see remarks before Chap. II, Definition 2.1),

$$\int_G F(x)\varphi(x)dx = \int_{G/H} \int_H F(xh)\varphi(xh)dh d\mu(\dot{x})$$

$$= \int_{G/H} f(\dot{x})T\varphi(\dot{x})d\mu(\dot{x}),$$

where $\dot{x} = xH$ and $T\varphi(\dot{x}) = \int_H \varphi(xh)dh$.

If $f = 0$ almost everywhere, then

$$\int_G F(x)\varphi(x)dx = 0$$

for all $\varphi \in C_c(G)$ and, hence, $F = 0$ almost everywhere.

Conversely, if $F = 0$ almost everywhere, then

$$\int_{G/H} f(\dot{x})T\varphi(\dot{x})d\mu(\dot{x}) = 0.$$

Hence, $f = 0$ almost everywhere, since $C_c(G/H) = \{T\varphi | \varphi \in C_c(G)\}$ (see [We], Chap. II, §9 or [Re], Chap. 3, 4.2). $\square$

2.9 Examples. Let G be a simple Lie group with finite centre. Let H be a non-compact closed subgroup, and let Γ be a lattice in G. Then Γ

acts ergodically on G/H. Indeed , by Moore's Ergodicity Theorem, H acts ergodically on G/Γ, and the conclusion follows from the Duality Theorem.

Here are two interesting examples:

(i) Consider the natural action of $\Gamma = \mathrm{SL}(n, \mathbb{Z})$ on $\mathbb{R}^n$. We may realize $\mathbb{R}^n \setminus \{0\}$ as the homogeneous space $\mathrm{SL}(n, \mathbb{R})/H$ where H denotes the stabilizer of the vector $(1, 0, \ldots, 0)^t$ in $\mathrm{SL}(n, \mathbb{R})$. Since H is not compact and since $\mathrm{SL}(n, \mathbb{Z})$ is a lattice in $\mathrm{SL}(n, \mathbb{R})$ (see Chap. V, Theorem 2.7), we conclude that the action of $\mathrm{SL}(n, \mathbb{Z})$ on $\mathbb{R}^n \setminus \{0\}$ and, hence, on $\mathbb{R}^n$ is ergodic. The same result holds for any lattice Γ in $\mathrm{SL}(n, \mathbb{R})$.

(ii) $\mathrm{SL}(2, \mathbb{R})$ acts in a natural way on the projective real line $\mathbb{R}P^1$. This (compact) space may be identified with $\mathrm{SL}(2, \mathbb{R})/P$, where P denotes the non-compact stabilizer of the point at infinity. As above, we conclude that any lattice Γ in $\mathrm{SL}(2, \mathbb{R})$ acts ergodically on $\mathbb{R}P^1$. It should be mentioned that $\mathbb{R}P^1$ has no non-zero Γ-invariant measure (see [Zi], Corollary 3.2.2).

§3 Counting Lattice Points in the Hyperbolic Plane

Let Γ be a lattice in $\mathrm{PSL}(2, \mathbb{R})$. We assume that Γ is torsion free. So, Γ acts freely on the hyperbolic plane $\mathbf{H}^2$ and $\Sigma = \Gamma \setminus \mathbf{H}^2$ is a surface of finite area.

Fix a point $p \in \mathbf{H}^2$, and let $B(p, R)$ denote the ball with centre p of radius R, with respect to the hyperbolic metric. The following is a classical lattice point problem:

Question: What can be said about the number $N(R, q)$ of points in an orbit Γq which lie in $B(p, R)$, for a point $q \in \mathbf{H}^2$?

We shall show that $N(R, q)$ has the asymptotic behaviour

$$N(R, q) \sim \frac{\mathrm{area}(B(p, R))}{\mathrm{area}(\Sigma)} \qquad (*)$$

as $R \to \infty$. (For real functions f and g, we write $f(t) \sim g(t)$ as $t \to \infty$ if $\lim_{t \to \infty} f(t)/g(t) = 1$.)

This is a special case of results established in Margulis' thesis [Ma4] and valid for manifolds with negative variable curvature. Following the proof appearing in [EM], we shall derive formula $(*)$ from the mixing of the geodesic flow, via an equidistribution theorem of spheres.

3.1 Exercise (Gauß' Circle Problem). The following analogue of the above problem in the Euclidean plane is much easier to prove. Let $N(R)$ be the number of lattice points (m, n) in $\mathbb{Z}^2$ with $m^2 + n^2 \leq R^2$. Show that

$$N(R) \sim \pi R^2$$

as $R \to \infty$. (Gauß' circle problem asks for good estimates for the error term.)

3.2 Lemma. *The area of a ball $B(p, R)$ in the hyperbolic plane* $\mathbf{H}^2$ *is*

$$\text{area}(B(p, R)) = 4\pi \sinh^2(R/2) = 2\pi(\cosh R - 1).$$

So, $\text{area}(B(p, R)) \sim \pi e^R$ *as* $R \to \infty$.

Proof This is best done by working in the disc model $\Delta = \{z \in \mathbb{C} | \, |z| < 1\}$ of the hyperbolic plane, with Riemannian metric (see Chap. II, Remark 1.9)

$$ds^2 = 2\frac{dx^2 + dy^2}{(1 - (x^2 + y^2))^2}.$$

The corresponding area element is

$$\left(\frac{2}{(1 - (x^2 + y^2))}\right)^2 dx dy.$$

One checks, as in the proof of Chap. II, Lemma 1.6, that

$$d(0, z) = \log\frac{1 + z}{1 - z}.$$

In particular, the hyperbolic ball $B(0, R) = \{z \in \Delta | \, d(0, z) \leq R\}$ is the Euclidean ball around 0 with radius $\tanh(R/2)$. Hence, the hyperpolic area of $B(0, R)$ is

$$\int_0^{\tanh(R/2)} \int_0^{2\pi} \left(\frac{2}{(1 - r^2)}\right)^2 r d\theta dr = 4\pi \sinh^2(R/2).$$

$\square$

Let d denote the hyperbolic metric on $\mathbf{H}^2$ as well as the metric it induces on $\Sigma = \Gamma \backslash \mathbf{H}^2$. For $x \in \Sigma$, let

$$S(x, t) := \{y \in \Sigma | \, d(x, y) = t\},$$

the sphere with centre x and radius t in Σ.

Applying an appropriate isometry if necessary, we may and shall assume throughout this section that $p = i$.

For a point $q \in \mathbf{H}^2$, let $[q]$ denote its image in Σ. Since $d(p, g_t \cdot p) = t$ (see Chap. II, Proposition 3.2), one has

$$S([p], t) = \{[k g_t \cdot p] | \, k \in K\},$$

where $g_t = \begin{pmatrix} e^{t/2} & 0 \\ 0 & e^{-t/2} \end{pmatrix}$ and $K = \text{SO}(2)$. Let λ_t be the measure on Σ supported on $S([p], t)$ and defined by integration with respect to the normalized Haar measure dk on K:

$$\int_\Sigma f d\lambda_t = \int_K f([kg_t \cdot p]) dk \qquad \forall f \in C_c(\Sigma).$$

3.3 Theorem. *The above measures λ_t are equidistributed, that is, for all $f \in C_c(\Sigma)$,*

$$\lim_{t\to\infty} \int_{S([p],t)} f d\lambda_t = \frac{1}{\text{area}(\Sigma)} \int_\Sigma f(x) d\mu(x),$$

where μ is the natural measure on Σ.

Proof Let $G = \text{SL}(2, \mathbb{R})$. We identify, as in Chap. II, Proposition 3.4, the unit tangent bundle $T^1(\Sigma)$ to $\Gamma \backslash G$ and Σ to $\Gamma \backslash G / K$.

Let $f \in C_c(\Gamma \backslash G / K)$, and let $\tilde{f}$ be the function f lifted to $\Gamma \backslash G$. Fix $\varepsilon > 0$. Choose a neighbourhood U of e in G with $KU \cap \Gamma = \{e\}$. This is possible since $K \cap \Gamma = \{e\}$. (Indeed, Γ is torsion free by assumption and, as K is compact, $K \cap \Gamma$ is a finite subgroup of Γ.) By uniform continuity of $\tilde{f}$, we may assume that

$$|\tilde{f}(gu) - \tilde{f}(g)| < \varepsilon, \qquad \forall g \in G, u \in U$$

Recall that G has an Iwasawa decomposition $G = K \cdot A \cdot N$ (see beginning of Chap. IV). Choose open neighbourhoods U_1 and U_2 of e in A and N such that $V := U_1 U_2 \subseteq U$.

Clearly (see the formulas in the proof of Chap. II, Lemma 3.5), shrinking U_2 if necessary, one has

$$g_t^{-1} U_2 g_t \subseteq U_2 \qquad \forall t \geq 0.$$

Hence, as $g_t \in A$ and A is abelian,

$$V g_t \subseteq g_t V \qquad \forall t \geq 0.$$

This implies

$$|\tilde{f}(\Gamma kvg_t) - \tilde{f}(\Gamma kg_t)| < \varepsilon \qquad \forall v \in V, k \in K,$$

and, hence,

$$\left| \int_K \tilde{f}(\Gamma kg_t) dk - \frac{1}{\mu(\Gamma KV)} \int_{\Gamma KV} \tilde{f}(\Gamma kvg_t) dk dv \right| < \varepsilon,$$

for all $t \geq 0$. Now, let χ be the characteristic function of the open subset ΓKV of $\Gamma \backslash G$. By Moore's Ergodicity Theorem (strong mixing of the geodesic flow; see Corollary 2.3),

$$\lim_{t\to\infty} \frac{1}{\mu(\Gamma KV)} \int_{\Gamma \backslash G} \tilde{f}(xg_t) \chi(x) dx = \lim_{t\to\infty} \frac{1}{\mu(\Gamma KV)} \int_{\Gamma KV} \tilde{f}(\Gamma kvg_t) dk dv$$

$$= \frac{1}{\mu(\Gamma \backslash G)} \int_{\Gamma \backslash G} \tilde{f}(x) dx.$$

Hence, for t large enough,

$$|\int_K f([kg_t \cdot p])dk - \frac{1}{\text{area}(\Sigma)}\int_\Sigma f(x)d\mu(x)| =$$

$$|\int_K \widetilde{f}(\Gamma kg_t)dk - \frac{1}{\mu(\Gamma \setminus G)}\int_{\Gamma\setminus G} \widetilde{f}(x)dx| < 2\varepsilon.$$

$\square$

3.4 Remark. Let φ_t be the geodesic flow on $T^1(\Sigma)$. The above set ΓKVg_t corresponds to a subset $\varphi_t(\Omega)$ of $T^1(\Sigma)$, where Ω consists of vectors lying over a small neighbourhood of $[p]$ and pointing in every possible direction. The set ΓKg_t corresponds to the set $\varphi_t(T_{[p]}(\Sigma))$ of vectors over points from the sphere $S([p], t)$ and normal to $S([p], t)$, and $\Gamma Kg_t U$ corresponds to a neighbourhood $\widetilde{\Omega}$ of $\varphi_t(T_{[p]}(\Sigma))$. A crucial point in the above proof was the fact that $\varphi_t(\Omega)$ is contained in $\widetilde{\Omega}$. Thus, $\varphi_t(\Omega)$ consists of vectors lying over points from a neighbouhood of $S(p, t)$ and nearly normal to $S(p, t)$. This is a feature of negative curvature. It is a special case of what is called "the wave front lemma" in [EM].

Recall that $N(R, q)$ is the number of points in an orbit Γq which lie in $B(p, R)$, for a point $q \in \mathbf{H}^2$.

3.5 Theorem. *For any* $q \in \mathbf{H}^2$, *one has*

$$N(R, q) \sim \frac{\text{area}(B(p, R))}{\text{area}(\Sigma)},$$

as R tends to infinity,

Proof Fix $q \in \mathbf{H}^2$, and denote by $[q]$ its image in Σ. Let $\varepsilon > 0$ be so small that $B(q, \varepsilon)$ is isometric to $B([q], \varepsilon)$.

Let α be a positive, continuous function on Σ with integral 1 and with support in $B([q], \varepsilon) \subseteq \Sigma$. Clearly, the function $x \mapsto N(R, x)$ lives on Σ. For any $x \in B(q, \varepsilon)$, one has

$$N(R - \varepsilon, q) \leq N(R, x) \leq N(R + \varepsilon, q).$$

Indeed, if $d(\gamma q, p) \leq R - \varepsilon$, for $\gamma \in \Gamma$, then

$$d(\gamma x, p) \leq d(\gamma x, \gamma q) + d(\gamma q, p) = d(x, q) + d(\gamma q, p) \leq R.$$

This proves the first inequality. The second one is shown in a similar way. Hence,

$$N(R - \varepsilon, q) \leq \int_\Sigma \alpha(x)N(R, x)dx \leq N(R + \varepsilon, q). \qquad (**)$$

Let $\widetilde{\alpha}$ denote the function α lifted to $\mathbf{H}^2 = G/K$. Then, denoting by χ

the characteristic function of $B(p, R)$,

$$\int_{\Sigma} \alpha(x) N(R, x) dx = \int_{\Sigma} \alpha(x) \sum_{\gamma \in \Gamma} \chi(\gamma x) dx = \int_{B(p,R)} \widetilde{\alpha}(x) dx.$$

Now (see next exercise),

$$\int_{B(p,R)} \widetilde{\alpha}(x) dx = 2\pi \int_0^R \int_K \widetilde{\alpha}(k g_t \cdot p) \sinh t \, dk \, dt$$

$$= 2\pi \int_0^R \left(\int_{\Sigma} \alpha d\lambda_t \right) \sinh t \, dt,$$

where λ_t is the measure on $S([p], R)$ defined above. By the previous equidistribution theorem we have, as α has integral 1,

$$\int_{\Sigma} \alpha d\lambda_t \to \frac{1}{\text{area}(\Sigma)} \qquad \text{as} \quad t \to \infty,$$

and, hence, as $\frac{1}{\text{area}(B(p,R))} \int_{B(p,R)} \widetilde{\alpha}(x) dx$ is a convex combination of such integrals,

$$\int_{B(p,R)} \widetilde{\alpha}(x) dx \sim \frac{\text{area}(B(p, R))}{\text{area}(\Sigma)} \qquad \text{as} \quad R \to \infty. \qquad (***)$$

Recall that $\text{area}(B(p, R)) \sim \pi e^R$ (see Lemma 3.2). Set

$$a(R) = \frac{N(R, q) \text{area}(\Sigma)}{\pi e^R}.$$

Then, by $(**)$,

$$e^{-\varepsilon} a(R - \varepsilon) \le \frac{\text{area}(\Sigma)}{\pi e^R} \int_{B(p,R)} \widetilde{\alpha}(x) dx \le e^{\varepsilon} a(R + \varepsilon).$$

Hence, if a is any limit point of $a(R)$ as $R \to \infty$, then, by $(***)$,

$$e^{-\varepsilon} a \le 1 \le e^{\varepsilon} a,$$

that is, $a \in [e^{-\varepsilon}, e^{\varepsilon}]$. Thus, since ε was arbitrary, $\lim_{R \to \infty} a(R)$ exists and is equal to 1. □

3.6 Exercise. Let (θ, t) be the coordinates on $\mathbf{H}^2$ defined by $z = k_\theta g_t \cdot i$, where k_θ is the rotation matrix of angle θ. Show that the area element $dx dy / y^2$ on $\mathbf{H}^2$ has the expression $2\pi \sinh t \, dt \, d\theta$ in the coordinates (θ, t).

Give another proof for the formula of the area of a ball of radius R (see Lemma 3.2).

§4 **Mixing of All Orders**

Let G be a locally compact group acting on a probability space (X, μ) in a measure preserving way. Strong mixing of this action means that any pair of measurable subsets A, B of X becomes independent at infinity, that is,

$$\lim_{g \to \infty} \mu(g \cdot A \cap B) = \mu(A)\mu(B), \qquad (g \in G)$$

or, equivalently, that

$$\lim_{n \to \infty} \mu(g_1(n) \cdot A \cap g_2(n) \cdot B) = \mu(A)\mu(B),$$

for all sequences $\{g_1(n)\}_{n \in \mathbb{N}}$ and $\{g_2(n)\}_{n \in \mathbb{N}}$ in G with

$$\lim_{n \to \infty} g_1(n)^{-1} g_2(n) = \infty.$$

It is natural to ask whether this phenomenom holds for more than two sets.

4.1 Definition. The action of G is called *k-mixing* ($k \geq 2$) if, for all measurable subsets $B_1, B_2, \ldots, B_k$ of X and all sequences

$$\{\overline{g}(n)\}_{n \in \mathbb{N}} = \{(g_1(n), g_2(n), \ldots, g_k(n))\}_{n \in \mathbb{N}} \in (G^k)^{\mathbb{N}}$$

with

$$\lim_{n \to \infty} g_i(n)^{-1} g_j(n) = \infty \qquad \forall \, 1 \leq i < j \leq k, \tag{1}$$

one has

$$\lim_{n \to \infty} \mu\left(\bigcap_{i=1}^{k} g_i(n) \cdot B_i \right) = \prod_{i=1}^{k} \mu(B_i). \tag{2}$$

Thus, strong mixing is 2-mixing. It is an old problem in ergodic theory whether a transformation (that is, an action of $\mathbb{Z}$) which is strongly mixing is mixing of all orders. Ledrappier [Ld] gave the following counterexample for an action of $\mathbb{Z}^2$.

Ledrappier's Counterexample

Let G be a group acting as a group of automorphisms on a compact abelian group X. Observe that the (normalized) Haar measure μ on X is automatically preserved by G (see the remarks before Chap. I, Proposition 1.5). Here is a criterion for strong mixing for the action of G on X.

4.2 Lemma. *Let G be an (infinite) discrete group acting by automorphisms on a compact abelian group X. Let $\widehat{X}$ be the dual group of X. Suppose that the dual action of G on $\widehat{X} \setminus \{1\}$ is free (that is, if the stabi-*

lizers of the characters $\chi \neq 1$ *are all trivial). Then the action of G on X is strongly mixing.*

Proof For $\chi_1, \chi_2 \in \widehat{X} \setminus \{1\}$, one has $\langle g \cdot \chi_1, \chi_2 \rangle \neq 0$ for at most one element $g \in G$, by the freeness assumption and by orthogonality of characters. Hence, if α and β are finite linear combinations of characters, then the matrix coefficient $g \mapsto \langle g \cdot \alpha, \beta \rangle$ has finite support. Since, by completeness of characters, such functions are dense in $L_0^2(X)$, this proves the claim. □

4.3 Remark. Actually, as is easy to prove, a stronger result is true: under the above assumption, the action of G has Lebesgue spectrum, that is, the associated representation of G on $L_0^2(X)$ is equivalent to a subrepresentation of a multiple of the regular representation of G.

The above criterion should be compared with the criterion for ergodicity which only requires that all orbits of G in $\widehat{X}$, except $\{1\}$, be infinite (see Chap. I, Proposition 1.5).

Let $Y := \{0, 1\}^{\mathbb{Z}^2}$ be the compact abelian group consisting of all sequences $\{y_{n,m}\}_{n,m \in \mathbb{Z}}$ of elements from the field $\mathbb{F}_2 = \{0, 1\}$, with the product topology. The group $\mathbb{Z}^2$ acts by automorphisms on Y by shifting the coordinates:

$$(k, l) \cdot \{y_{n,m}\}_{n,m} = \{y_{n+k,m+l}\}_{n,m} \qquad \forall (k, l) \in \mathbb{Z}^2.$$

The Haar measure on Y is $\mu = \otimes_{i \in \mathbb{Z}^2} \mu_i$, where $\mu_i(\{0\}) = \mu_i(\{1\}) = 1/2$, for all $i \in \mathbb{Z}^2$.

Let X be the closed subgroup of Y consisting of the *harmonic sequences*, that is, the sequences $\{y_{n,m}\}_{n,m}$ such that

$$y_{n,m} = y_{n-1,m} + y_{n,m-1} + y_{n+1,m} + y_{n,m+1} \qquad (*)$$

for all $n, m \in \mathbb{Z}$. (Of course, the identity $(*)$ is to be understood as an equality in $\mathbb{F}_2$.)

Clearly, X is invariant under the action of $\mathbb{Z}^2$. This gives rise to an action of $\mathbb{Z}^2$ by automorphisms of X. We claim that the action of $\mathbb{Z}^2$ on X obtained this way is strongly mixing but is not mixing of all orders.

4.4 Proposition. *The above action of $\mathbb{Z}^2$ on X is not 5-mixing.*

Proof Define $E := \{y \in X \mid y_{0,0} = 1\}$ and let

$$g_1(n) = (0, 0), \ g_2(n) = (2^n, 0), \ g_3(n) = (0, 2^n),$$
$$g_4(n) = (-2^n, 0), \ g_5(n) = (0, -2^n).$$

Then $g_i(n)^{-1}g_j(n)$ tends to ∞ for $i < j$ as $n \to \infty$, and

$$\bigcap_{i=1}^{5} g_i(n) \cdot E = \{y \in X \mid y_{0,0} = y_{2^n,0} = y_{0,2^n} = y_{-2^n,0} = y_{0,-2^n} = 1\}.$$

We claim that $\bigcap_{i=1}^{5} g_i(n) \cdot E$ is empty. Indeed, by induction, $(*)$ implies that

$$y_{n,m} = y_{n-2^k,m} + y_{n,m-2^k} + y_{n+2^k,m} + y_{n,m+2^k}$$

for all $k \in \mathbb{N}_0$. Hence, if $y \in \bigcap_{i=1}^{5} g_i(n) \cdot E$, then,

$$1 = y_{0,0} = y_{2^n,0} + y_{0,2^n} + y_{-2^n,0} + y_{0,-2^n} = 0,$$

a contradiction. $\square$

Let $\widehat{X}$ be the dual group of X. We are going to show that the action of $\mathbb{Z}^2$ on $\widehat{X} \setminus \{1\}$ is free. By the above Lemma 4.2 , this will prove our claim. First, we parametrize $\widehat{X}$ by pairs of finite subsets of $\mathbb{Z}$. Let $P_f(\mathbb{Z})$ be the set of all finite subsets of $\mathbb{Z}$. For a pair (F, G) in $P_f(\mathbb{Z}) \times P_f(\mathbb{Z})$, let $\sigma_{F,G}$ be the character on X defined by

$$\sigma_{F,G}(y) = \prod_{n \in F} (-1)^{y_{n,0}} \prod_{m \in G} (-1)^{y_{m,1}},$$

for all $y = \{y_{n,m}\}_{n,m} \in X$.

4.5 Lemma. *The mapping*

$$(F, G) \mapsto \sigma_{F,G}$$

is a bijection between $P_f(\mathbb{Z}) \times P_f(\mathbb{Z})$ *and the dual group* $\widehat{X}$.

Proof By standard character theory (see [HeR], Chap. VI), any character of X is the restriction of a character of Y. Clearly, any σ from $\widehat{Y}$ is of the form $\sigma = \sigma_A$ for a uniquely determined finite subset A of $\mathbb{Z}^2$, where

$$\sigma_A(y) = \prod_{(n,m) \in A} (-1)^{y_{n,m}}, \qquad \forall y = \{y_{n,m}\}_{n,m} \in Y.$$

Now, any $y = \{y_{n,m}\}_{n,m}$ in X is determined by the two sequences $\{y_{n,0}\}_n$ and $\{y_{m,1}\}_m$. Moreover, for any two sequences $\{a_n\}_n$ and $\{b_m\}_m$ in $\{0,1\}^{\mathbb{Z}}$, there exists a $y = \{y_{n,m}\}_{n,m}$ in X with

$$\{y_{n,0}\}_n = \{a_n\}_n \quad \text{and} \quad \{y_{m,1}\}_m = \{b_m\}_m.$$

This shows that two characters σ_A and σ_B from $\widehat{Y}$ agree on X if and only if the subsets A and B of $\mathbb{Z}^2$ agree on each of the two lines $\{(n,0) \mid n \in \mathbb{Z}\}$ and $\{(m,1) \mid m \in \mathbb{Z}\}$. $\square$

It will be convenient to identify subsets of $\mathbb{Z}$ with elements from the group algebra $\mathbb{F}_2[\mathbb{Z}]$ of $\mathbb{Z}$ over $\mathbb{F}_2$: a subset A of $\mathbb{Z}$ is identified with its characteristic function $\chi_A \in \mathbb{F}_2[\mathbb{Z}]$. The group algebra $\mathbb{F}_2[\mathbb{Z}]$ is a commutative algebra over $\mathbb{F}_2$ with convolution product and $1 = \delta_0$, the characteristic function of 0, as unit. The convolution product of $f, g \in \mathbb{F}_2$ will be denoted by $f * g$. We record a few elementary properties of $\mathbb{F}_2[\mathbb{Z}]$.

4.6 Lemma. *Let A and B be subsets of $\mathbb{Z}$, and let $f = \chi_A$ and $g = \chi_B$, viewed as elements in $\mathbb{F}_2[\mathbb{Z}]$. Then*
(i) $\delta_n * f = \chi_{n+A}$, for all $n \in \mathbb{Z}$;
(ii) $f + g = \chi_{A \triangle B}$, where $A \triangle B$ denotes the symmetric difference;
(iii) If $f * g = 0$, then $f = 0$ or $g = 0$ (that is, $\mathbb{F}_2[\mathbb{Z}]$ has no divisor of zero).

Next, we describe the action of $\mathbb{Z}^2$ on $\widehat{X}$ when $\widehat{X}$ is identified with $\mathbb{F}_2[\mathbb{Z}] \oplus \mathbb{F}_2[\mathbb{Z}]$, by means of Lemma 4.5 above.

4.7 Lemma. *The action of $\mathbb{Z}^2$ on $\widehat{X} \cong \mathbb{F}_2[\mathbb{Z}] \oplus \mathbb{F}_2[\mathbb{Z}]$ is generated by the transformations T, S defined by*

$$T(f,g) = \begin{pmatrix} \delta_1 & 0 \\ 0 & \delta_1 \end{pmatrix} \begin{pmatrix} f \\ g \end{pmatrix} = (\delta_1 * f, \delta_1 * g)$$

$$S(f,g) = \begin{pmatrix} 0 & 1 \\ 1 & \varrho \end{pmatrix} \begin{pmatrix} f \\ g \end{pmatrix} = (g, f + \varrho * g),$$

where $\varrho = \chi_{\{-1,0,1\}} = \delta_{-1} + \delta_0 + \delta_1$.

Proof The action of $\mathbb{Z}^2$ on $\widehat{X}$ is generated by the transformations T, S corresponding to the action of $(1,0)$ and $(0,1)$. For any pair (A,B) of finite subsets A and B of $\mathbb{Z}$, one has $T(A,B) = (A+1, B+1)$. So, T has the matrix $\begin{pmatrix} \delta_1 & 0 \\ 0 & \delta_1 \end{pmatrix}$, when viewed as a transformation on $\mathbb{F}_2[\mathbb{Z}] \oplus \mathbb{F}_2[\mathbb{Z}]$.

The value of $S(A,B)$ on $y = \{y_{n,m}\}_{n,m} \in X$ is given by

$$S(A,B)y = \prod_{n \in A} (-1)^{y_{n,1}} \prod_{m \in B} (-1)^{y_{m,2}}.$$

Since

$$y_{m,2} = y_{m,1} + y_{m-1,1} + y_{m+1,1} + y_{m,0},$$

one has

$$S(A,B)y = \prod_{m \in B} (-1)^{y_{m,0}} \prod_{n \in A, m \in B} (-1)^{y_{m,1}+y_{m-1,1}+y_{m+1,1}+y_{n,1}}$$

$$= \prod_{m \in B} (-1)^{y_{m,0}} \prod_{k \in C} (-1)^{y_{k,1}},$$

where $C = A \triangle (B-1) \triangle (B+1) \triangle B$. Since, by Lemma 4.6,

$$\chi_C = f + \chi_{\{-1,0,1\}} * g$$

for $f = \chi_A$ and $g = \chi_B$, this proves the claim. $\square$

4.8 Lemma. *Let $n \in \mathbb{N}$, $n \geq 2$. Then*

$$S^n = \begin{pmatrix} a_n & b_n \\ b_n & a_n + \varrho * b_n \end{pmatrix},$$

where $a_n, b_n \in \mathbb{F}_2[\mathbb{Z}]$ are polynomials in ϱ of degree $n-2$ and $n-1$, respectively. Moreover, the coefficient of ϱ^{n-2} in b_n is 0.

Proof Writing $S^{n+1} = S^n \begin{pmatrix} 0 & 1 \\ 1 & \varrho \end{pmatrix}$, this follows immediately by induction on n. $\square$

4.9 Lemma. *Let $p, q \in \mathbb{Z}$, $q \geq 0$, not both zero. Then $T^p S^q(f,g) \neq (f,g)$ for all $(f,g) \in \mathbb{F}_2[\mathbb{Z}] \oplus \mathbb{F}_2[\mathbb{Z}]$, with $(f,g) \neq (0,0)$.*

Proof With the notations of Lemma 4.7, for $m \in \mathbb{Z}$, $n \in \mathbb{N}$,

$$T^m S^n = \begin{pmatrix} \delta_m a_n & \delta_m b_n \\ \delta_m b_n & \delta_m a_n + \delta_m \varrho b_n \end{pmatrix},$$

where we denote by fg the convolution product $f * g$. Suppose that the linear homogeneous system $T^p S^q(f,g) = (f,g)$ has a non-zero solution. Then, necessarily,

$$(\delta_p a_q - 1)(\delta_p a_q + \delta_p \varrho b_q - 1) - \delta_p^2 b_q^2 = 0.$$

As (f,g) is also a fixed point of $T^{kq} S^{kq} = (T^q, S^q)^k$, one even has

$$\delta_{2kp} a_{kq}^2 + \delta_{2kp} b_{kq}^2 + \delta_{2kp} \varrho a_{kq} b_{kq} + \delta_{kp} \varrho b_{kq} + 1 = 0$$

for all $k \in \mathbb{Z}$, that is,

$$a_{kq}^2(2kp+n) + b_{kq}^2(2kp+n) + \varrho a_{kq} b_{kq}(2kp+n) + \varrho b_{kq}(kp+n) + \delta_0(n) = 0 \quad (**)$$

for all $k, n \in \mathbb{Z}$. Observe that, by the previous lemma, a_{kq} and b_{kq} are polynomials in ϱ of degree $kq - 2$ and $kq - 1$.

- *First case:* suppose that $p > q$. Then

$$a_q^2(2p) = \varrho a_q b_q(2p) = \varrho b_q(p) = b_q^2(2p) = 0.$$

Taking $k = 1$ and $n = 0$ in $(**)$ yields a contradiction.

- *Second case:* suppose that $p < q$. Fix $k \geq 3$, and let $n = 2(kq - 1) - 2kp$, so that $2kp + n = 2(kq - 1)$. Then $b_{kq}^2(2kp + n) = 1$ and $a_{kq}^2(2kp + n) = \varrho b_{kq} a_{kq}(2kp + n) = 0$. As $q - p > 0$ and $k \geq 3$,

$$|kp + n| = |kq + k(q - p) - 2| > kq.$$

Hence, $\varrho b_{kq}(kp+n) = 0$ since ϱb_{kq} has degree kq, and $(**)$ yields a contradiction.

• *Third case:* suppose that $p = q$. Take $k = 1$ and $n = -1$ in $(**)$. On the one hand,

$$a_q^2(2p-1) = b_q^2(2p-1) = \varrho a_q b_q(2p-1) = 0.$$

On the other hand, since the coefficient of $\varrho^{p-2} = \varrho^{q-2}$ in b_q is 0 (see previous lemma), $\varrho b_q(p-1) = \varrho^p(p-1) + 0 = 1$. This is a contradiction to $(**)$. $\qquad\square$

4.10 Corollary. *The action of $\mathbb{Z}^2$ on $\widehat{X} \setminus \{1\}$ is free. Hence, the action of $\mathbb{Z}^2$ on X is strongly mixing.*

Proof By the previous lemma, it remains to show that $T^p S^q$ has no fixed point $(f,g) \neq (0,0)$ if $q < 0$. This is clear, since $(T^p S^q)^{-1} = T^{-p} S^{-q}$ has the same fixed points as $T^p S^q$. $\qquad\square$

Mixing of All Orders for Actions of Semisimple Groups

Mozes proved that ergodic actions of semisimple Lie groups are mixing of all orders (see [Mz1], [Mz2]). We shall give his proof in the case of the action of a simple Lie group on a compact measure space (see Remark 4.15 below)

4.11 Theorem. *Let (X,μ) be a compact, metrizable probability space and G be a simple non-compact Lie group with finite centre. If G acts on X in a continuous, measure-preserving and ergodic way, then G is mixing of all orders.*

Observe that, since mixing of all orders is inherited by closed subgroups, the theorem, combined with Chap. II, Corollary 3.8, implies that the geodesic flow of a Riemann surface of constant negative curvature is mixing of all orders. This is also true for the horocycle flow, to be discussed in the next chapter.

We first establish some general lemmas.

4.12 Lemma. *Let G be a connected semisimple Lie group, with centre Z. Then $\mathrm{Ad}(G)$ is the connected component of the group of all Lie algebra automorphisms of $\mathfrak{g}$. In particular, $\mathrm{Ad}(G)$ is closed in $\mathrm{GL}(\mathfrak{g})$, and $\mathrm{Ad}(G)$ and G/Z are isomorphic as topological groups.*

Proof The Lie algebra of the automorphisms of $\mathfrak{g}$ can be identified with that of all derivations on $\mathfrak{g}$. But every derivation of a semisimple Lie algebra is inner (see [Hel], Chap. II, 6.4), showing the claim. $\qquad\square$

4.13 Lemma. *Let H be a locally compact group acting continuously on a compact metric space Y. Let $\{\nu_n\}_n$ be a sequence of probability measures on Y and $\{h_n\}_n$ a sequence in H such that ν_n is h_n -invariant. Assume that $\lim_n \nu_n = \nu$ in the weak-$*$ topology and $\lim_n h_n = h$, for some probability measure ν on Y and some $h \in H$. Then ν is h -invariant.*

Proof Let $\varphi \in C(Y)$, and let $\varepsilon > 0$. Since H acts continuously and since φ is uniformly continuous, there is $n_1 \in \mathbb{N}$ such that for all $n \geq n_1$,

$$|\varphi(h_n y) - \varphi(hy)| < \varepsilon \quad \forall y \in Y.$$

Since $\lim_n \nu_n = \nu$, there exists n_2 with

$$\left| \int_Y \varphi(hy)d\nu(y) - \int_Y \varphi(hy)d\nu_n(y) \right| < \varepsilon$$

$$\left| \int_Y \varphi(y)d\nu(y) - \int_Y \varphi(y)d\nu_n(y) \right| < \varepsilon$$

for all $n \geq n_2$. Then, for $n \geq \max\{n_1, n_2\}$, using the h_n -invariance of ν_n,

$$\left| \int_Y \varphi(hy)d\nu(y) - \int_Y \varphi(y)d\nu(y) \right| \leq \left| \int_Y \varphi(hy)d\nu(y) - \int_Y \varphi(hy)d\nu_n(y) \right|$$

$$+ \left| \int_Y \varphi(hy)d\nu_n(y) - \int_Y \varphi(h_n y)d\nu_n(y) \right|$$

$$+ \left| \int_Y \varphi(y)d\nu_n(y) - \int_Y \varphi(y)d\nu(y) \right|$$

$$< 3\varepsilon.$$

$\qquad\square$

4.14 Lemma. *Let (Y, ν) and (Z, η) be probability spaces. Let ω be a probability measure on $Y \times Z$ such that its projections on Y and Z are ν and η . Let H be a group with an ergodic action on (Z, η). Assume that ω is invariant under the action of $\mathrm{id} \times H$. Then $\omega = \mu \otimes \eta$.*

Proof We desintegrate ω with respect to the projection $\pi : Y \times Z \to Y$ (see [Bol], §3, Théorème 1): there exists a measurable mapping $y \mapsto \omega_y$ from Y to $\mathcal{M}^1(Z)$, with $\operatorname{supp}\omega_y \subseteq \pi^{-1}(y)$ and such that

$$\int_{Y \times Z} \varphi d\omega = \int_Y \left(\int_Z \varphi d\omega_y \right) d\nu,$$

for all $\varphi \in L^1(Y \times Z, \omega)$, since $\pi(\omega) = \nu$. We view the measures ω_y as measures on Z. Since the projection of ω on Z is η, we have

$$\eta = \int_Y \omega_y d\nu(y). \qquad (***)$$

Now, $\mathrm{id} \times H$ preserves ω. Hence, by unicity of desintegration, ν-almost all measures ω_y are H-invariant. But $(***)$ is a decomposition of the H-ergodic measure η as an average of H-invariant measures. Hence, by extremality of ergodic measures (see Chap. I, Proposition 3.1), $\eta = \omega_y$ for ν-almost all y, showing that $\omega = \nu \otimes \eta$. $\qquad \Box$

Proof of Theorem 4.11. Let $k \geq 2$, and let Δ be the diagonal in X^k, identified with X. Let $\mu_{\Delta,k}$ be the image of the measure μ under this identification, that is,

$$\int_{X^k} \varphi(x_1, \ldots, x_k) d\mu_{\Delta,k} = \int_X \varphi(x, \ldots, x) d\mu$$

for all $\varphi \in C(X^k)$. Let $\{\bar{g}(n)\}_{n \in \mathbb{N}} = \{(g_1(n), g_2(n), \ldots g_k(n))\}_{n \in \mathbb{N}}$ be a sequence in G^k with

$$\lim_{n \to \infty} g_i(n)^{-1} g_j(n) = \infty \qquad \forall 1 \leq i < j \leq k. \qquad (1)$$

We have to show that

$$\lim_{n \to \infty} \bar{g}(n) \cdot \mu_{\Delta,k} = \mu^{\otimes k}, \qquad (2)$$

where $\mu^{\otimes k}$ is the product measure on X^k and where

$$\int_{X^k} \varphi(x_1, \ldots, x_k) d\bar{g} \cdot \nu := \int_{X^k} \varphi(g_1^{-1} x_1, \ldots, g_k^{-1} x_k) d\nu$$

for $\bar{g} = (g_1, \ldots, g_k) \in G^k$, $\nu \in \mathcal{M}(X^k)$ and $\varphi \in C(X^k)$.

Since X is compact and metrizable, the weak-$*$-topology on the space of all probability measures $\mathcal{M}^1(X^k)$ on X^k is compact and metrizable with respect to the weak-$*$ topology. Thus, in order to show (2), it suffices to show that $\mu^{\otimes k}$ is the unique weak-$*$ accumulation point of the sequence $\{\bar{g}(n) \cdot \mu_{\Delta,k}\}_{n \in \mathbb{N}}$.

So, let ω be such a weak-$*$ accumulation point. We have to prove that

$$\omega = \mu^{\otimes k}. \qquad (3)$$

This will be achieved by showing ω is invariant under a non-trivial unipotent element of G^k. We proceed by induction on the mixing order $k \geq 2$.

For $k = 2$, the claim follows from the Howe–Moore Vanishing Theorem (see Theorem 1.1).

We assume from now on that $k > 2$. First of all, we may assume that $g_1(n) = e$ for all $n \in \mathbb{N}$. Indeed, $\bar{f}(n) = (e, g_1(n)^{-1} g_2(n), \ldots, g_1(n)^{-1} g_k(n))$

satisfies (1) and , for $\varphi \in C(X^k)$,

$$\int_{X^k} \varphi d\overline{f}(n)\mu_{\Delta,k} = \int_X \varphi(x, g_2(n)^{-1}g_1(n)x, \ldots, g_k(n)^{-1}g_1(n)x)d\mu(x)$$

$$= \int_X \varphi(g_1(n)^{-1}x, g_2(n)x^{-1}x, \ldots, g_k(n)^{-1}x)d\mu(x)$$

$$= \int_{X^k} \varphi d\overline{g}(n)\mu_{\Delta,k},$$

by the G-invariance of μ.

Let $\|\cdot\|$ be any norm on the Lie algebra $\mathfrak{g}$ of G. Choose $\varepsilon > 0$ so that the exponential mapping $\exp: B(0,\varepsilon) \to \exp(B(0,\varepsilon))$ is a diffeomorphism. There exists $b \in B(0,\varepsilon)$ with $\|b\| \geq \varepsilon/2$ such that

$$\lim_{n\to\infty} \|\mathrm{Ad}(g_2(n))b\| = \infty.$$

Indeed, $\lim_{n\to\infty} g_2(n) = \infty$ (by (1)). Hence, as $\mathrm{Ad}(G)$ is homeomorphic to the quotient of G by its finite centre (see Lemma 4.12 above),

$$\lim_{n\to\infty} \mathrm{Ad}(g_2(n)) = \infty$$

in $\mathrm{GL}(\mathfrak{g})$. Since $\mathrm{Ad}(G)$ is semisimple,

$$\det \mathrm{Ad}(g_2(n)) = 1 \qquad \forall n \in \mathbb{N}.$$

It follows that $\lim_{n\to\infty} \mathrm{Ad}(g_2(n)) = \infty$ in $\mathrm{End}(\mathfrak{g})$, the space of all linear mappings from $\mathfrak{g}$ to itself. Thus, there exists $a \in \mathfrak{g} \setminus \{0\}$ such that

$$\lim_{n\to\infty} \|\mathrm{Ad}(g_2(n))a\| = \infty.$$

Now, set

$$b := \frac{\varepsilon}{2\|a\|}a.$$

Since $\lim_{n\to\infty} \mathrm{Ad}(g_2(n)b) = \infty$, by continuity of the adjoint action, there exists a sequence of real numbers $\{t(n)\}_{n\in\mathbb{N}}$ such that

$$\max_{1\leq j\leq k} \|\mathrm{Ad}(g_j(n))t(n)b\| = \varepsilon. \tag{4}$$

Let $\{v(n)\}_{n\in\mathbb{N}}$ be the sequence in G^k defined by

$$v(n) := \big(\exp(t(n)b), \exp(\mathrm{Ad}(g_2(n))t(n)b), \ldots, \exp(\mathrm{Ad}(g_k(n))t(n)b)\big).$$

Let $d(\cdot,\cdot)$ be a left invariant metric on G, and let

$$0 < \alpha := \min_{\|a\|=\varepsilon} d(\exp(a), e), \qquad \beta := \max_{\|a\|=\varepsilon} d(\exp(a), e).$$

Then $\{v(n)\}_{n\in\mathbb{N}}$ is contained in the compact set

$$M := \{\overline{r} = (r_1, \ldots, r_k) \in G^k \mid \alpha \leq \max_{1\leq j\leq k} d(r_j, e) \leq \beta\}.$$

Hence, upon passing to a subsequence, we may assume that

$$\overline{h} = (h_1, h_2, \ldots, h_k) = \lim_{n \to \infty} v(n)$$

exists. Then $h_1 = e$. Reorder the coordinates to achieve

$$h_1 = \cdots = h_\ell = e, \qquad \text{and} \qquad h_j \neq e \quad \forall j > \ell.$$

Observe that $h \neq e$, since $e \notin M$. Hence, $\ell < k$.

We claim that the measure ω on X^k is h–invariant. Indeed, as ω is a limit point of the measures $\overline{g}(n) \cdot \mu_{\Delta, k}$ which are clearly invariant under $v(n)$, this follows from Lemma 4.13.

We claim that each $\mathrm{Ad}(h_j)$ is unipotent for $\ell < j \leq k$, that is, there exists $m \in \mathbb{N}$ such that $(\mathrm{id} - \mathrm{Ad}(h_j))^m = 0$.

Indeed, on the one hand,

$$h_j = \lim_{n \to \infty} \exp(\mathrm{Ad}(g_j(n))t(n)b) = \lim_{n \to \infty} g_j(n) \exp(t(n)b) g_j(n)^{-1}.$$

On the other hand, observe that, by (4), $\lim_{n \to \infty} t(n) = 0$, since

$$\lim_{n \to \infty} \|\mathrm{Ad}\, g_2(n)b\| = \infty.$$

Hence,

$$\lim_{n \to \infty} \exp(t(n)b) = e.$$

Therefore, $\mathrm{Ad}(h_j)$ has the same characteristic polynomial as the identity matrix. Thus, 1 is the unique eigenvalue of $\mathrm{Ad}(h_j)$, and the claim follows.

To finish the proof, decompose

$$X^k = Y \times Z,$$

where Y is the product of the first ℓ factors, and Z is the product of the remaining ones. Since $\ell \notin \{0, k\}$, the induction hypothesis applies to both factors. Hence, the projections ν, η of ω on Y and Z are the product measures $\mu^{\otimes \ell}$ and $\mu^{\otimes k - \ell}$, respectively.

Now the closed subgroup H_j of G generated by h_j, $j > \ell$, is not compact. Indeed, otherwise, the unipotent and non-trivial matrix $\mathrm{Ad}(h_j)$ would be contained in a compact group. Since compact groups act in a semisimple way, $\mathrm{Ad}(h_j)$ would be diagonalizable, and this is impossible.

By Moore's Ergodicity Theorem, H_j is strongly mixing on X. Hence, $(h_{\ell+1}, \ldots, h_k)$ acts ergodically on $(Z, \mu^{\otimes k - \ell})$. Now Lemma 4.14 applies and shows that

$$\omega = \mu^{\otimes \ell} \otimes \mu^{\otimes k - \ell} = \mu^{\otimes k}.$$

$\square$

4.15 Remark. The above theorem extends immediately to actions of a simple Lie group on general Lebesgue spaces. This requires the use of some general measure theoretic results of Ramsay and Varadarajan. Moreover, as the proof shows, the result remains true for strongly mixing actions of so–called Ad -proper Lie groups on Lebesgue spaces (see [Mz1], [Mz2]).

Notes

As demonstrated in Vilenkin's monograph [Vi], many classical special functions arise as matrix coefficients of groups representations. Due of course to Harish–Chandra's work (see [Wr1], [Wr2]), the study of matrix coefficients of semisimple Lie groups is now a highly developed subject. More precise information is known about their asymptotic behaviour than just vanishing at infinity (see [BW], [CaM], [Co], [How]). This is obtained through detailed estimates using the fine structure of the groups. For instance, it is known that the matrix coefficients of a non-trivial irreducible representation of a simple real (or even p -adic) Lie group are in some L^p .

Asymptotics of matrix coefficients have been studied for other groups. In fact, one of the main concerns of Howe and Moore's article [HM] was to establish a vanishing result for arbitrary algebraic groups over local fields. A description of the connected groups for which all non-trivial irreducible representations have vanishing matrix coefficients is given in [Sch2].

Let $X = G/K$ be a Riemannian symmetric space of rank 1. Let $V = \Gamma \setminus X$ be a compact quotient of X by a discrete group of isometries acting freely on X. The geodesic flow of V is strongly mixing (see Corollary 2.3). This means that all matrix coefficients $t \mapsto \langle T_{\exp tH} f, g \rangle$ of the associated unitary representation on $L_0^2(T^1 V)$ of the one-parameter subgroup $A = \exp \mathfrak{a}$ of G vanish at infinity ($f, g \in L_0^2(T^1 V)$). An important problem is to obtain more precise information about the rate of mixing of the geodesic flow, that is, about the decay of these matrix coefficients at infinity.

Burger and Schroeder [BuS] give estimates for the rate of mixing of the geodesic flow when f and g are K -fixed functions, using the asymptotic behaviour of spherical functions on G/K, and derive a nice application (an "isoperimetric inequality") of this result to the geometry of V. Estimates for more general matrix coefficients are given in [Mo3] and [Ra11] for $G = \mathrm{SL}(2, \mathbb{R})$ and in [KM1] for arbitrary semisimple Lie groups.

For an account on ergodic properties of flows on nilmanifolds (these are manifolds of the type $\Gamma \setminus G$ for a nilpotent Lie group G and a lattice Γ in G), see [Pr1], [Pr2]. Further results on ergodicity of flows may be found in [BM].

In the case of a cocompact lattice Γ, the asymptotic formula for the number of lattice points in the hyperpolic plane (Theorem 3.5) was proved by Delsarte ([De1], [De2]) and Huber ([Hub]). The case of a non-cocompact lattice was considered by Patterson ([Pat]). Their proofs use an expansion of the counting function with respect to the eigenfunctions of the Laplace operator on $\mathbf{H}^2/\Gamma$ and allow more precise results. Counting problems have also been studied in the *infinite* volume case (see, for instance, [BP], [PhR], [PS] and [Ro]). In [EM] and [EMS1], counting results are proved for symmetric (not necessarily Riemannian) spaces (see also [DRS], [Bar], [Gün] and [LP]). In [Ma4], an asymptotic formula is given for manifolds with variable negative curvature.

Ledrappier's example (Proposition 4.4) motivated the deep study of $\mathbb{Z}^d$-actions by automorphisms on a compact abelian group made by K. Schmidt in [Sch1].

Marcus [Mr] showed that horocycle flows (see Chapter IV) are mixing of all orders and conjectured the same to be true for general ergodic actions of semisimple Lie groups. As seen above, this is indeed the case by Mozes' result (*loc. cit.*). Ornstein and Weiss [OW] proved that the geodesic flow of a Riemann surface of constant negative curvature is a Bernoulli flow, a property implying mixing of all orders. This was generalized by Dani [Da10] to actions of semisimple elements on a finite volume homogeneous space. Using Ratner's measure classification theorem (see Chapter VI), Starkov [St2] proved mixing of all orders for all mixing flows on a finite volume homogeneous space.

Chapter IV

The Horocycle Flow

In this chapter, we discuss some basic results on the horocycle flow of finite area surfaces covered by the hyperbolic plane. It is described in group theoretical terms in Section 1. Its mixing is then an immediate consequence of Moore's Ergodicity Theorem.

The dynamical behaviour of the orbits of the geodesic flow is very complicated. For example, it is known that there are many non-periodic and non-dense orbits in that case (see [Mor]). By way of contrast, the horocycle flow has a much rigid behaviour. For instance, every orbit is either dense or periodic (Hedlund's theorem). In Section 2, we give the proof of Hedlund's theorem in the case of a cocompact lattice.

Section 3 and Section 4 are devoted to stronger results: the classification of all ergodic measures with respect to the horocycle flow (proved by Furstenberg [Fu1] in the cocompact case, and by Dani [Da1] in general) and the equidistribution theorem of orbits of the horocycle flow (due to Dani–Smillie [DS]). These are now a special case of Ratner's results (see the Notes in Chapter VI). We follow the proofs from [Rat7] which contain some ingredients of her methods for the general case.

Throughout this chapter, we use the following subgroups of $SL(2, \mathbb{R})$:

$$N = \left\{ u(x) := \begin{pmatrix} 1 & x \\ 0 & 1 \end{pmatrix} \mid x \in \mathbb{R} \right\}$$

$$A = \left\{ a(\tau) := \begin{pmatrix} e^\tau & 0 \\ 0 & e^{-\tau} \end{pmatrix} \mid \tau \in \mathbb{R} \right\}$$

$$N^- = \left\{ h(x) := \begin{pmatrix} 1 & 0 \\ x & 1 \end{pmatrix} \mid x \in \mathbb{R} \right\}$$

$$P = A \cdot N = \left\{ \begin{pmatrix} e^\tau & x \\ 0 & e^{-\tau} \end{pmatrix} \mid x, \tau \in \mathbb{R} \right\}$$

as well as $K = SO(2)$. One has the *Iwasawa decomposition*

$$SL(2, \mathbb{R}) = KAN$$

and the *Bruhat decomposition*

$$SL(2, \mathbb{R}) = N^- AN \cup \omega AN, \qquad \omega = \begin{pmatrix} 0 & -1 \\ 1 & 0 \end{pmatrix}.$$

Explicitly,

$$\begin{pmatrix} a & b \\ c & d \end{pmatrix} = \begin{pmatrix} \cos\theta & -\sin\theta \\ \sin\theta & \cos\theta \end{pmatrix} \begin{pmatrix} \alpha & 0 \\ 0 & \alpha^{-1} \end{pmatrix} \begin{pmatrix} 1 & \beta \\ 0 & 1 \end{pmatrix},$$

$$\alpha = \sqrt{a^2 + c^2}, \qquad \cos\theta = \frac{a}{\sqrt{a^2 + c^2}}, \qquad \beta = \frac{ab + cd}{\sqrt{a^2 + c^2}}$$

and

$$\begin{pmatrix} a & b \\ c & d \end{pmatrix} = \begin{pmatrix} 1 & 0 \\ c/a & 1 \end{pmatrix} \begin{pmatrix} a & 0 \\ 0 & a^{-1} \end{pmatrix} \begin{pmatrix} 1 & b/a \\ 0 & 1 \end{pmatrix}, \qquad \text{if } a \neq 0$$

$$\begin{pmatrix} a & b \\ c & d \end{pmatrix} = \begin{pmatrix} 0 & -1 \\ 1 & 0 \end{pmatrix} \begin{pmatrix} c & 0 \\ 0 & c^{-1} \end{pmatrix} \begin{pmatrix} 1 & d/c \\ 0 & 1 \end{pmatrix}, \qquad \text{if } a = 0.$$

In particular, observe that $N^- AN$ is an open and dense subset of $\text{SL}(2, \mathbb{R})$.

§1 The Horocycle Flow of a Riemann Surface

Consider again the Poincaré half plane

$$\mathbf{H}^2 = \{z \in \mathbb{C} \mid \text{Im} z > 0\}$$

with the Riemannian structure discussed in Section II.1. Observe that any hyperbolic circle in $\mathbf{H}^2$ is a Euclidean circle and vice versa. Indeed, if C is the hyperbolic circle $\{z \in \mathbf{H}^2 \mid d(z, z_0) = r\}$ around $z_0 = x_0 + iy_0 \in \mathbf{H}^2$, then, by the formula in Chap. II, Lemma 1.7,

$$C = \{z \in \mathbf{H}^2 \mid |z - w_0| = \rho\},$$

where

$$w_0 = x_0 + (2R^2 + 1)y_0, \quad R = \sinh(r/2) \quad \text{and} \quad \rho = 2y_0 R \sqrt{R^2 + 1}.$$

This also shows that, conversely, an Euclidean circle is a hyperbolic circle.

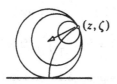

Horocycle in the Poincaré half plane

Let $(z, \zeta) \in T^1 \mathbf{H}^2$, and let σ be the unique geodesic determined by (z, ζ). Thus, $\sigma : \mathbb{R} \to \mathbf{H}^2$ is the curve with $\sigma(0) = z$ and $\dot{\sigma}(0) = \zeta$. For $t \in \mathbb{R}$, let C_t be the hyperbolic circle with centre $\sigma(t)$ and passing through z. When $t \to +\infty$ (respectively $t \to -\infty$), C_t converges to a circle tangent to the x-axis or to a line parallel to the x-axis and having ζ as inward (respectively outward) normal vector at z.

This "limit circle" $C_{\pm\infty}$ is the *positive (respectively negative) horocycle* determined by (z, ζ). We define a flow on $T^1 \mathbf{H}^2$ as follows.

Let z_t be the point on the horocycle $C_{+\infty}$ at a distance t from z (positive orientation of the circle). Let ζ_t be the unit inward normal vector to $C_{+\infty}$ at z_t (see figure). We set $\psi_t^+ (z, \zeta) = (z_t, \zeta_t)$. The one-parameter group

$$\psi_t^+ : T^1 \mathbf{H}^2 \to T^1 \mathbf{H}^2$$

is called the *positive horocycle flow*.

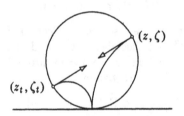

Horocycle flow in the Poincaré half plane

Similarly, one defines the negative horocycle flow

$$\psi_t^- : T^1 \mathbf{H}^2 \to T^1 \mathbf{H}^2,$$

using the negative horocycles. The positive and the negative horocycle flows are conjugated by the symmetry of $T^1 \mathbf{H}^2$ taking (p, ζ) into the opposite line element $(p, -\zeta)$.

1.1 Theorem. *Under the identification $T^1 \mathbf{H}^2 \cong \mathrm{PSL}(2, \mathbb{R})$, the horocycle flow ψ_t^+ (respectively ψ_t^-) corresponds to the right translations*

$$\mathrm{PSL}(2, \mathbb{R}) \to \mathrm{PSL}(2, \mathbb{R}) \qquad g \mapsto g\, u(t) \quad \textit{resp.} \quad g \mapsto g\, h(t),$$

where $u(t) = \begin{pmatrix} 1 & t \\ 0 & 1 \end{pmatrix}$, $h(t) = \begin{pmatrix} 1 & 0 \\ t & 1 \end{pmatrix}$, for all $t \in \mathbb{R}$.

Proof Using Chap. II, Lemma 3.1, we see that it suffices to consider the pair (z_0, ζ_0), where $z_0 = i$ and ζ_0 is the unit tangent vector at z_0

pointing in the positive direction of the imaginary axis. For $(z_t^+, \zeta_t^+) = \psi_t^+(z_0, \zeta_0)$, one clearly has $z_t^+ = t + i$, $\zeta_t^+ = \zeta_0$. Therefore, $(z_t^+, \zeta_t^+) = u(t) \cdot (z_0, \zeta_0)$. Similarly, $\psi_t^-(z_0, \zeta_0) = h(t) \cdot (z_0, \zeta_0)$.

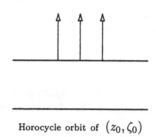

Horocycle orbit of (z_0, ζ_0)

$\square$

1.2 Remark. The horocycle flow is connected to the geodesic flow via

$$\varphi_s \circ \psi_t^\pm \circ \varphi_{-s} = \psi_t^\pm \exp{(\pm s)}, \qquad \forall t, s \in \mathbb{R},$$

where φ_s is the geodesic flow. This is a consequence of the fundamental formulae

$$a(\tau)u(t)a(\tau)^{-1} = u(te^{2\tau}), \qquad a(\tau)h(t)a(\tau)^{-1} = h(te^{-2\tau}). \qquad (1)$$

So, the geodesic flow normalizes the horocycle flow. It contracts the positive and expands the negative horocycles.

Let $\Sigma = \Gamma \backslash \mathbf{H}^2$ for a lattice Γ in $\mathrm{SL}(2, \mathbb{R})$. As we know (see Chap. II, Proposition 3.4), the unit tangent bundle $T^1\Sigma$ may be identified with $\Gamma \backslash T^1\mathbf{H}^2$. The horocycle flow on $T^1\mathbf{H}^2$ may be pushed down to a flow on $T^1\Sigma$, the horocycle flow on $T^1\Sigma$. Under the identification $T^1\Sigma \cong \Gamma \backslash \mathrm{PSL}(2, \mathbb{R})$, these flows correspond to the right translations

$$\Gamma g \mapsto \Gamma g\, u(t) \quad (\text{resp. } \Gamma g \mapsto \Gamma g\, h(t)).$$

Since N and N^- are non-compact, Moore's Ergodicity Theorem (see Chap. III, Theorem 2.1) implies that the horocycle flow is mixing, a result due to Hedlund [He2].

1.3 Theorem. *The horocycle flow of the finite volume surface $\Gamma \backslash \mathbf{H}^2$ is strongly mixing (and hence ergodic).*

1.4 Remark. The ergodicity of the horocycle flow follows in a more direct way from Chap. III, Theorem 1.6.

Hedlund's Minimality Theorem

We are going to study the horocycle flow from a topological point of view, that is, we investigate its dynamical properties.

Recall that the action of the group G on the topological space X is called *minimal* if all the orbits $G \cdot x$, $x \in X$, are dense in X. Equivalently, the only closed G-invariant subsets of X are X and $\emptyset$. The following proposition shows that an ergodic action already has some minimality properties.

1.5 Proposition. *Let G be a locally compact group acting continuously on a second countable topological space X. Assume that X is equipped with a Borel measure μ such that $\mu(U) > 0$ for any non-empty open $U \subseteq X$ and that μ is invariant and ergodic under the action of G.*

(i) *Let S be a subsemigroup of G such that $G = S \cup S^{-1}$. Then, for almost any $x \in X$, the orbit Sx is dense in G;*

(ii) *Assume that μ is a probability measure and that the action of G on X is strongly mixing. Let $\{s_i\}_{i \in \mathbb{N}}$ be a sequence in G with $\lim s_i = \infty$. Then, for almost any $x \in X$, the sequence $\{s_i x\}_{i \in \mathbb{N}}$ is dense in G.*

Proof (i) Let U be an open non-empty set in X. Set

$$A = \bigcup_{s \in S} s^{-1}(U).$$

Clearly, $s^{-1}A \subseteq A$ for all $s \in S$. Hence,

$$s^{-1}A \subseteq A \subseteq sA, \qquad \forall s \in S.$$

Moreover, $\mu(A) > 0$, as A is open. As $\mu(s^{-1}A) = \mu(A) = \mu(sA)$ for all $s \in S$ and $G = S \cup S^{-1}$, the characteristic function χ_A of A is essentially G-invariant. By ergodicity, it follows that χ_A is constant almost everywhere (see Chap. I, Theorem 1.3). Hence, $\mu(X \setminus A) = 0$.

Let $\{U_n\}_{n \in \mathbb{N}}$ be a (countable) basis for the topology on X. By what we have just seen,

$$\mu\left(X \setminus \left(\bigcap_{n \in \mathbb{N}} \bigcup_{s \in S} s^{-1}(U_n)\right)\right) = 0.$$

Now, the S-orbit of any point $x \in \bigcap_{n \in \mathbb{N}} \bigcup_{s \in S} s^{-1}(U_n)$ is dense. Indeed, $Sx \cap U_n \neq \emptyset$ for all n, and hence $Sx \cap U \neq \emptyset$ for any open non-empty set U.

(ii) Let U be an open non-empty subset of X, and let

$$A = \bigcup_{i \in \mathbb{N}} s_i^{-1}(U).$$

Then

$$\mu(U) = \mu(s_i^{-1}(U) \cap A) = \langle \chi_{s_i^{-1}(U)}, \chi_A \rangle.$$

By strong mixing (see Chap. I, Remark 2.11 (i)),

$$\langle \chi_{s_i^{-1}(U)}, \chi_A \rangle \to \mu(U)\mu(A), \quad \text{as} \quad i \to \infty.$$

Hence, $\mu(A) = 1$, since $\mu(U) > 0$. The claim follows as in (i). $\qquad\square$

Observe that the above proposition applies to $S = G$. So, for almost any $x \in X$, the orbit Gx is dense, when the action of G is ergodic. We need the more general statement (i) above for the proof of Lemma 2.4 below.

From now on, we shall restrict to the positive horocycle flow ψ_t^+. We start with a description of the periodic orbits. For this we need to recall some notions from Chap. II, §2. Let Γ be a lattice in $\mathrm{SL}(2, \mathbb{R})$. The cusps of Γ are the Γ-orbits of fixed points of the parabolic elements in Γ. We identify these cusps with vertices $x_1, x_2, \ldots, x_n$ at infinity of a Dirichlet region D for Γ. Let $X = \Gamma \backslash \mathrm{SL}(2, \mathbb{R})$.

1.6 Proposition. *Let $x_1, x_2, \ldots, x_n \in \mathbb{R} \cup \{\infty\}$ be the cusps of Γ, and $g_1, g_2, \ldots, g_n \in \mathrm{SL}(2, \mathbb{R})$ be such that $x_i = g_i \cdot \infty$, $1 \le i \le n$. Then $x = \Gamma g \in X$ has a periodic orbit under the horocycle flow if and only if there exist a $\gamma \in \Gamma$, an upper triangular matrix $p \in P$ and an i_0, $1 \le i_0 \le n$, such that $\gamma g = g_{i_0} p$.*

Proof Each x_i is a fixed point of a parabolic element $\gamma_i \in \Gamma$. Thus the eigenvalues of γ_i are both 1 (or both -1). In addition,

$$g_i^{-1} \gamma_i g_i \cdot \infty = g_i^{-1} \gamma_i \cdot x_i = g_i^{-1} \cdot x_i = \infty.$$

Thus, $g_i^{-1} \gamma_i g_i$ is an upper triangular matrix, hence of the form $u(x)$, $x \in \mathbb{R}$. On the other hand, all fixed points of parabolic elements of Γ are of the form $\gamma \cdot x_i$ for suitable $\gamma \in \Gamma$ and $1 \le i \le n$.

Now consider a point $x = \Gamma g \in X$ with a periodic orbit. Then, $\Gamma g u(t) = \Gamma g$ for some $t > 0$. Hence, we find $\gamma \in \Gamma$ with $g u(t) = \gamma g$. So, $\gamma = g u(t) g^{-1}$ is a parabolic element in Γ fixing $g \cdot \infty$. Therefore, there exist $\gamma' \in \Gamma$ and $i_0 \in \{1, \ldots, n\}$ such that

$$g \cdot \infty = \gamma' \cdot x_{i_0} = \gamma' g_{i_0} \cdot \infty,$$

that is, $g^{-1} \gamma' g_{i_0} \cdot \infty = \infty$. This implies that $g_{i_0}^{-1} \gamma'^{-1} g$ is an upper triangular matrix p, and $\gamma'^{-1} g = g_{i_0} p$.

Conversely, let $g = g_i p$ for an upper triangular matrix $p = a(\tau)u(x)$. Using formula (1) and the fact that $g_i^{-1}\gamma_i g_i = u(t_0)$ for a certain $t_0 \neq 0$, one has

$$\Gamma g_i p u(t) = \Gamma g_i (pu(t)p^{-1})p = \Gamma g_i u(te^{2\tau})p$$
$$= \Gamma \gamma_i g_i u(te^{2\tau})p = \Gamma g_i u(t_0 + te^{2\tau})p$$
$$= \Gamma g_i p u(t_0 e^{-2\tau} + t).$$

Hence, $T := t_0 e^{-2\tau}$ is a non-zero period of $\Gamma g_i p$. $\qquad\square$

1.7 Example. Let $\Gamma = \mathrm{SL}(2,\mathbb{Z})$. The points in $X = \Gamma \backslash \mathrm{SL}(2,\mathbb{R})$ with periodic orbits under the horocycle flow are the points Γp for $p \in P$.

If Γ is a cocompact lattice in $\mathrm{PSL}(2,\mathbb{R})$, then a Dirichlet region for the action of Γ on $\mathbf{H}^2$ has no vertex at infinity. Hence, the previous proposition has the following consequence.

1.8 Corollary. *The horocycle flow of* $\Gamma \backslash \mathrm{SL}(2,\mathbb{R})$ *has no periodic orbit if* Γ *is cocompact.*

1.9 Theorem (Hedlund). *Let* Γ *be a lattice in* $\mathrm{PSL}(2,\mathbb{R})$. *Any orbit of the horocycle flow on* $\Gamma \backslash \mathrm{PSL}(2,\mathbb{R})$ *is either dense or periodic. In particular, the horocycle flow is minimal if* Γ *is cocompact.*

We shall first treat the case of a cocompact lattice for which the proof is easier. The case of a general lattice will be considered later in Section 3, as a consequence of a more general result.

§2 Proof of Hedlund's Theorem – Cocompact Case

General Properties of Minimal Invariant Sets

First, we have to establish some elementary facts which will also be used in Chapter VI for the proof of Oppenheim's conjecture.

2.1 Lemma. *Let* G *be a locally compact group acting continuously on a compact space* Ω. *Then there exists a closed,* G*-invariant subset* $Y \neq \emptyset$ *of* Ω *which is minimal with respect to these properties.*

Proof Let

$$\mathcal{O} := \{A \subseteq \Omega \mid A \text{ is closed, non-empty and } G\text{-invariant}\}.$$

Then $\mathcal{O}$ is not empty and inductively ordered by inclusion. Indeed, if $\mathcal{F}$ is a totally ordered family in $\mathcal{O}$, the finite intersection property implies that

$$\bigcap_{A \in \mathcal{F}} A$$

is non-empty, and is a lower bound of $\mathcal{F}$ in $\mathcal{O}$. By Zorn's lemma, $\mathcal{O}$ contains minimal elements. □

2.2 Lemma. *Let G be a locally compact group acting continuously on a topological space Ω. Let A, B and C be closed subgroups of G, where C is contained in $A \cap B$. Let X and Y be closed subsets of Ω. Suppose that X is invariant under A, that Y is invariant under B, and that Y is compact and minimal under the action of C. Let M be a subset of G with*

$$mY \cap X \neq \emptyset, \qquad \forall m \in M.$$

Then

$$gY \subseteq X, \qquad \forall g \in N_G(C) \cap \overline{AMB},$$

where $N_G(C)$ denotes the normalizer of C in G.
Hence, if in addition $A = B$ and $X = Y$, then

$$gY = Y \qquad \forall g \in N_G(C) \cap \overline{AMA}.$$

Proof Clearly, $gY \cap X \neq \emptyset$ for every $g \in AMB$. By continuity and compactness of Y, the same is true for every $g \in \overline{AMB}$. If, in addition, g normalizes C, then, for every $y \in Y$ with $gy \in X$ and for all $c \in C$,

$$gcy = (gcg^{-1})gy \subseteq CX = X.$$

Now, by minimality of Y under C, one has $Y = \overline{Cy}$, and hence, $gY \subseteq X$.

As to the second statement, observe that the first part shows that gY is a non-empty closed subset of Y which is C-invariant, since $g \in N_G(C)$. The claim follows now from the minimality of Y. □

2.3 Lemma. *Let G be a locally compact group, and let Ω be a homogeneous space of G. Let C be a closed subgroup of G and let Y be a closed subset of Ω. Assume that Y is C-invariant and minimal under the action of C. If there exists $y \in Y$ with $Y \neq Cy$, that is, if Y is not a C-orbit, then the unit element $e \in G$ is in the closure of the set*

$$M := \{g \notin C \mid gy \in Cy\}.$$

Proof Indeed, assume by contradiction, that $e \notin \overline{M}$. Then, we find a neighbourhood V of e with the property that, if $u \in V$ is such that $uy \in Cy$, then $u \in C$. Choose an open neighboourhood U of e such that $U^2 \subseteq V$.

We claim that

$$Uy \cap Y = Uy \cap Cy = (U \cap C)y.$$

Indeed, let $u \in U$ with $uy \in Y$. Then, as $Y = \overline{Cy}$, there exists $u_i \in U$ with $\lim_i u_i = e$ such that $u_i uy \in Cy$. As $u_i u \in U^2 \subseteq V$, one has $u_i u \in C$. Hence, $u = \lim_i u_i u \in C$, proving the claim.

So, we have $CUy \cap Y = CUy \cap Cy$. Observe that the non-empty set $CUy \cap Y$ is C-invariant and open in Y. Hence, by minimality,

$$Y = CUy \cap Y = CUy \cap Cy \subseteq Cy \subseteq Y.$$

So, $Y = Cy$, a contradiction.   □

Recall that $\mathrm{PSL}(2, \mathbb{R})$ acts (by fractional linear transformations) transitively on the boundary

$$B = \mathbb{R} \cup \{\infty\} \cong \mathbb{PR}^1$$

of $\mathbf{H}^2$. The stabilizer of the point ∞ is the subgroup P of the upper triangular matrices. So, B may be identified with the homogeneous space $\mathrm{PSL}(2, \mathbb{R})/P$, the so-called *Furstenberg boundary* of $\mathrm{PSL}(2, \mathbb{R})$. The following is a classical fact.

2.4 Lemma. *Let* Γ *be a lattice in* $G = \mathrm{SL}(2, \mathbb{R})$. *The action of* Γ *on* $B = G/P$ *is minimal. Equivalently, the action of* P *on* $\Gamma \backslash G$ *is minimal.*

Proof (Mostow)   We have to show that, for any $x \in G = \mathrm{SL}(2, \mathbb{R})$, the double coset $\Gamma x P$ is dense in G.

Since $\Gamma x P = x (x^{-1} \Gamma x) P$, and since $x^{-1} \Gamma x$ is a lattice, it is sufficient to show that $\overline{\Gamma P} = G$. So, let $g \in G$. We know that the action of A on $\Gamma \backslash G$ is ergodic (see Chap. II, Corollary 3.8) Hence, by (i) of the above Proposition 1.5,

$$\Gamma u (A^+)^{-1}$$

is dense in G for almost all $u \in G$ and therefore for u in a dense subset. So, we may find sequences $\{\gamma_n\}_n \subseteq \Gamma$, $\{a_n\}_n \subseteq A^+$ and $\{u_n\}_n$ in the open dense subset $N^- P$ such that $u_n \to e$ and $\gamma_n u_n^{-1} a_n^{-1} \to g$ as $n \to \infty$. Set $w_n := \gamma_n u_n^{-1} a_n^{-1} g^{-1}$. Then $\gamma_n = w_n g a_n u_n$ and $w_n \to e$. Write $u_n = v_n p_n$ with $v_n \in N^-$, $p_n \in P$. Then, as $n \to \infty$,

$$\gamma_n P = w_n g a_n v_n P = w_n g (a_n v_n a_n^{-1}) a_n P \to gP,$$

since $w_n \to e$ and $a_n v_n a_n^{-1} \to e$ and $a_n \in P$.   □

2.5 Remark.   The above proof is valid, without any change, for any semisimple Lie group G with finite centre and without compact factors and for $B = G/P$ where P is a parabolic subgroup (see [Mos], Lemma 8.5).

Proof of Hedlund's Theorem – Cocompact Case

Let $G = \mathrm{SL}(2,\mathbb{R})$ and assume that $\Gamma \backslash G$ is compact. By Lemma 2.1, there exists a minimal closed non-empty N-invariant subset X of $\Gamma \backslash G$. We claim that $X = \Gamma \backslash G$.

Since there are no periodic N-orbits (Corollary 1.8), Lemma 2.3 implies, in particular, that

$$e \in \overline{M}, \qquad (*)$$

where $M := \{g \notin N \mid Xg \cap X \neq \emptyset\}$. By Lemma 2.2,

$$X = Xg \qquad \forall g \in P \cap \overline{NMN}, \qquad (**)$$

since P normalizes N. By the above minimality lemma (Lemma 2.4), it suffices to show that P is contained in $\overline{NMN}$.

Clearly, $M = NMN$ and $N \cup M = \{g \in G \mid Xg \cap X \neq \emptyset\}$ is a closed subset of G, since X is compact. Hence, $\overline{M} \subseteq N \cup M$. Therefore, by $(*)$,

$$\overline{M} = \overline{NMN} = N \cup M. \qquad (***)$$

Consider now the natural action of $\mathrm{SL}(2,\mathbb{R})$ on $\mathbb{R}^2 \backslash \{0\}$. The N-orbits in $\mathbb{R}^2 \backslash \{0\}$ are

- the lines which are parallel to the x-axis and
- the points on the x-axis.

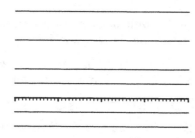

The orbits of N

Let

$$\Phi : \mathrm{SL}(2,\mathbb{R}) \to \mathbb{R}^2 \backslash \{0\}, \quad g \mapsto ge_1, \quad e_1 = \begin{pmatrix} 1 \\ 0 \end{pmatrix}.$$

The stabilizer of e_1 is N. So, Φ induces an identification of $\mathrm{SL}(2,\mathbb{R})/N$ with $\mathbb{R}^2 \backslash \{0\}$.

The image $M' = \Phi(\overline{M})$ of $\overline{M}$ is a closed subset M' of $\mathbb{R}^2 \backslash \{0\}$, since Φ is an open mapping. Moreover, M' is a a union of N-orbits, as $\overline{M} = \overline{NMN} = N\overline{M}$. Since $\overline{M} = N \cup M$ is a disjoint union, one has $\Phi(M) = M' \backslash \{e_1\}$, and it follows from $(*)$ that e_1 is in the closure of $M' \backslash \{e_1\}$. Hence, either M' contains a sequence of horizontal lines converging to the x-axis (in which case M' contains the whole x-axis) or M' contains a sequence of points lying on the x-axis, different from

e_1 and converging to e_1. In both cases, there exists a sequence $\{m'_i\}_i \in M' \setminus \{e_1\}$ lying on the x-axis such that $\lim_i m'_i = e_1$.

For each i, let $m_i \in M$ be such that $\Phi(m_i) = m'_i$. Then $m_i \in P$, since $m'_i = m_i e_1$ lies on the x-axis and $m_i \notin N$, since $m'_i \neq e_1$. Observe that, by (∗∗) and (∗∗∗),

$$P \cap \overline{M} = \{g \in P \mid X = Xg\}.$$

Hence, $P \cap \overline{M}$ is a closed subgroup of P containing N. For the images $[m_i]$ in $(P \cap \overline{M})/N$ of the m_i's under the canonical projection, one has

$$[m_i] \to e \qquad \text{and} \qquad [m_i] \neq e,$$

where e is the group unit of $(P \cap \overline{M})/N$. Hence, $(P \cap \overline{M})/N$ is a non-discrete closed subgroup of P/N. As $P/N \cong \mathbb{R}$, it follows that

$$(P \cap \overline{M})/N = P/N,$$

that is, P is contained in $\overline{M}$.

2.6 Remark. The last step in the above proof is a special case of a general fact about orbits of unipotent one-parameter groups acting on a vector space. The more general statement will be given in Chapter VI and used for the proof of Oppenheim's conjecture (see Chap. VI, Lemma 4.1).

We conclude this section with an application of Hedlund's theorem (cocompact case) to quadratic forms in two variables.

2.7 Exercise. Let $f : \mathbb{R}^2 \to \mathbb{R}$ be a bilinear form in two variables. Let Γ be a cocompact lattice in $\mathrm{SL}(2, \mathbb{R})$. Prove that the set

$$\{f(\gamma_{11}, \gamma_{21}) \mid \gamma = (\gamma_{ij})_{1 \leq i,j \leq 2} \in \Gamma\}$$

is dense in the set $f(\mathbb{R}^2)$ of all values of f. (Observe also that one cannot remove the cocompactness condition on Γ, as already the example $\Gamma = \mathrm{SL}(2, \mathbb{Z})$ and $f(x, y) = x$ shows.)

Hint: Consider the Γ-orbit of $e_1 = (1, 0)^t$, under the natural action of $\mathrm{SL}(2, \mathbb{R})$ on $\mathbb{R}^2$.

§3 Classification of Invariant Measures

We now give the classification of the invariant measures under the horocycle flow. Observe that, by the decomposition theorem (Chap. I, Theorem 3.2), it suffices to classify the ergodic measures.

3.1 Theorem (Classification of Ergodic Measures). *Let* Γ *be a lattice in* $G := \mathrm{SL}(2, \mathbb{R})$ *and let* μ *be a probability measure on* $X := \Gamma \backslash G$ *which is invariant and ergodic under the horocycle flow. Then* μ *is either* G – *invariant (and thus* μ *is the unique* G –*invariant probability measure* ν_G *on* X *) or* μ *is supported by a periodic orbit.*

When Γ is cocompact, there is no periodic orbit (see Corollary 1.8). Hence, we have the following result due to H. Furstenberg [Fu1].

3.2 Corollary. *For a cocompact lattice* Γ *in* $\mathrm{SL}(2, \mathbb{R})$, *the horocycle flow is uniquely ergodic.*

In its full generality, the classification result was first proved by Dani [Da1]. We shall follow the proof given by Ratner [Ra7].

For a continuous function f on G or on $X = \Gamma \backslash G$ and x in G or X, define the averages of f along the orbit xN

$$S_f(x, t) := \frac{1}{t} \int_0^t f(xu(s))\, ds, \quad u(s) = \begin{pmatrix} 1 & s \\ 0 & 1 \end{pmatrix}, \quad t > 0.$$

3.3 Proposition. *Assume that, for every* $x \in X$ *with a non-periodic* N-*orbit and for every* $f \in C_c(X)$, *there exists a sequence* $\{t_n\}_n$ *of real numbers with* $t_n \to \infty$, *such that*

$$\lim_{n \to \infty} S_f(x, t_n) = \int_X f\, d\nu_G. \tag{*}$$

Then the Haar measure ν_G *is the only ergodic* N-*invariant probability measure which is not supported on a periodic orbit.*

Proof Let μ be an N-ergodic probability measure on X. Let $f \in C_c(X)$. Birkhoff's ergodic theorem (see Chap. I, Theorem 2.5) shows that, for μ-almost all $x \in X$,

$$\lim_{t \to \infty} S_f(x, t) = \int_X f\, d\mu. \tag{**}$$

Let $Y = \bigcup_{i=1}^n \Gamma g_i P$ be the set of points with N-periodic orbits (see Proposition 1.6). Clearly, Y is measurable and N-invariant. Since μ is ergodic, $\mu(Y) = 1$ or $\mu(Y) = 0$. In the first case, as the periodic orbits are compact, μ is supported by a periodic orbit, by Chap. I, Theorem 2.20. In the second case, $\mu(Y) = 0$ and it follows from (*) and (**) that $\mu = \nu_G$. $\qquad\square$

For given $x \in X$ with a non-periodic N-orbit and $f \in C_c(X)$, we shall construct a sequence t_n as above. As a corollary, we obtain the proof of Hedlund's minimality result in full generality .

Proof of 1.9 (general case) Let $x \in X$ with a non-periodic orbit. Let Ω be an open non-empty subset of X. Let $f \in C_c(X)$ with support contained in Ω and with non-zero integral. It is now obvious from equality $(*)$ above that $xN \cap \Omega \neq \emptyset$. $\square$

Proof of the Classification Theorem

First of all, recall the notation

$$u(s) = \begin{pmatrix} 1 & s \\ 0 & 1 \end{pmatrix} \in N, \quad a(\tau) = \begin{pmatrix} e^{\tau} & 0 \\ 0 & e^{-\tau} \end{pmatrix} \in A, \quad h(t) = \begin{pmatrix} 1 & 0 \\ t & 1 \end{pmatrix} \in N^{-}$$

and the relations

$$\begin{aligned} a(\tau)u(t)a(-\tau) &= u(te^{2\tau}) \\ a(\tau)h(t)a(-\tau) &= h(te^{-2\tau}). \end{aligned} \tag{1}$$

For $x \in G$ and for y close to x, we want to study the divergence of the N-orbit of Γy from the N-orbit of Γx.

Recall that $N^{-}AN$ is an open and dense subset of $G = \mathrm{SL}(2, \mathbb{R})$. For small $\delta > 0$, define

$$\mathbf{W}(x, \delta) := \{xa(\tau)h(s) | \; |\tau|, |s| < \delta\}.$$

For every $y = xa(\tau)h(r) \in \mathbf{W}(x, \delta)$, and every $s \in [0, 1]$, one has

$$yu(\Theta(y, s)) \in \mathbf{W}(xu(s), 10\delta)$$

for

$$\Theta(y, s) = \frac{s}{e^{2\tau} - s\tau}. \tag{2}$$

Indeed, using the relations (2), one verifies that

$$yu(\Theta(y, s)) = xu(s)a(\tau_s)h(r_s),$$

and $|\tau_s|, |r_s| < 10\delta$, where $\tau_s = \ln(e^{\tau} - sre^{-\tau})$, $r_s = r(1 - sre^{-2\tau})$.

3.4 Lemma. *For every $\varepsilon > 0$, there is a $\delta(\varepsilon) > 0$ such that, for all $0 < \delta < \delta(\varepsilon)$,*

$$\left| \frac{\partial}{\partial s} \Theta(y, s) - 1 \right| < \varepsilon,$$

for all $y \in \mathbf{W}(x, \delta)$ and all $0 \leq s \leq 1$.

Proof By formula (2),

$$\frac{\partial}{\partial s} \Theta(y, s) = \frac{e^{2\tau}}{(e^{2\tau} - s\tau)^2},$$

and the claim is clear. $\square$

Fix a large $t > 0$, and let $\tau = (\ln t)/2$. Define

$$\mathbf{W}(x, \delta, t) := \mathbf{W}(xa(\tau), \delta)a(-\tau) = \{xa(\xi)h(r)|\ |\xi| < \delta, |r| < \delta e^{-2\tau} = \delta/t\}.$$

For $y = xa(\xi)h(r) \in \mathbf{W}(x, \delta, t)$, one has

$$ya(\tau) = xa(\tau)a(\xi)h(rt) \in \mathbf{W}(xa(\tau), \delta).$$

Define

$$\Theta(y, s) := \Theta(ya(\tau), \frac{s}{t})t = \frac{s}{e^{2\xi} - \frac{s}{t}r}. \tag{3}$$

Then,

$$yu(\Theta(y, s)) = ya(\tau)u(\Theta(ya(\tau), \frac{s}{t}))a(-\tau) \in \mathbf{W}(xu(s), 10\delta, t).$$

3.5 Lemma. *For every* $0 < \varepsilon < 1/2$, *there exists* $\delta(\varepsilon) > 0$ *such that, for all* $0 < \delta < \delta(\varepsilon)$,

$$\left| \frac{\partial}{\partial s} \Theta(y, s) - 1 \right| < \varepsilon, \tag{4}$$

for all $y \in \mathbf{W}(x, \delta, t)$, *all* $0 < s \le t$ *and all* $t > 0$. *In particular,*

$$\left| \frac{t}{\Theta(y, t)} - 1 \right| < 2\varepsilon, \tag{5}$$

for all $y \in \mathbf{W}(x, \delta, t)$, *and all* $t > 0$.

Proof By formula (3),

$$\frac{\partial}{\partial s} \Theta(y, s) = \frac{e^{2\xi}}{(e^{2\xi} - \frac{s}{t}r)^2},$$

and the claim follows from the previous lemma, with the same $\delta(\varepsilon)$. Integrating gives

$$\left| \frac{\Theta(y, t)}{t} - 1 \right| < \varepsilon, \qquad \forall t > 0,$$

from which inequality (5) immediately follows. $\qquad\qquad\qquad\square$

Now, let f be a *right* uniformly continuous bounded function on G, that is, for every $\varepsilon > 0$, there exists a neighbourhood $U(\varepsilon)$ of e such that

$$|f(y) - f(z)| < \varepsilon, \qquad \forall y, z \in G, \ y \in zU(\varepsilon). \tag{6}$$

Recall that, for $y \in G$,

$$S_f(y, t) := \frac{1}{t} \int_0^t f(yu(s))\, ds.$$

3.6 Lemma. *For every* $0 < \varepsilon < 1/2$, *there exists* $\delta(\varepsilon, f) > 0$ *such that*

$$|S_f(y, \Theta(y, t)) - S_f(x, t)| < \varepsilon, \tag{7}$$

for all $\delta \le \delta(\varepsilon, f)$, *all* $y \in \mathbf{W}(x, \delta, t)$ *and all* $t \ge 1$.

Proof Let $\tilde{\varepsilon} = \varepsilon/(\|f\|_\infty + 1)$, and let $\tilde{\delta}$ be such that

$$\{a(\xi)h(r)|\ |\xi|, |r| < \delta\} \subseteq U(\varepsilon), \qquad \forall\, 0 < \delta \leq \tilde{\delta},$$

where $U(\varepsilon)$ is a neighbourhood of e as above. Let $\delta(\tilde{\varepsilon})$ be such that (4) holds . Let

$$\delta(\varepsilon, f) = \frac{1}{10}\min\{\tilde{\delta}, \delta(\tilde{\varepsilon})\}.$$

Let $\delta \leq \delta(\varepsilon, f)$, $t \geq 1$ and $y \in \mathbf{W}(x, \delta, t)$. Then, by change of variable,

$$S_f(y, \Theta(y,t)) = \frac{1}{\Theta(y,t)} \int_0^t f(yu(\Theta(y,s)))\frac{\partial}{\partial s}\Theta(y,s)ds.$$

Now,

$$yu(\Theta(y,s)) \in \mathbf{W}(xu(s), 10\delta, t) \subseteq \mathbf{W}(xu(s), \tilde{\delta}).$$

Hence, using (4), (6) and (5), one has, for all $y \in \mathbf{W}(x, \delta, t)$ and all $t \geq 1$,

$$|S_f(y, \Theta(y,t)) - S_f(x,t)| \leq \frac{1}{\Theta(y,t)} \int_0^t |f(yu(\Theta(y,s)))(\frac{\partial}{\partial s}\Theta(y,s) - 1)|ds$$

$$+ \frac{1}{\Theta(y,t)} \int_0^t |f(yu(\Theta(y,s))) - f(xu(s))|ds$$

$$+ |\frac{1}{\Theta(y,t)} - \frac{1}{t}| \int_0^t |f(xu(s))|ds$$

$$\leq 2\frac{t}{\Theta(y,t)}\varepsilon + |\frac{t}{\Theta(y,t)} - 1|\|f\|_\infty$$

$$\leq 2(1 + 2\tilde{\varepsilon})\varepsilon + 2\varepsilon \leq 8\varepsilon.$$

$$\square$$

In order to deal with non-cocompact lattices, we need one fact about A-orbits of points with a non periodic N-orbit.

First, we recall some facts from Chapter II about the structure of fundamental domains for Fuchsian groups. Let Γ be a (non-cocompact) lattice in $G = \mathrm{SL}(2, \mathbb{R})$. Let $x_1, x_2, \ldots, x_n \in \mathbb{R} \cup \{\infty\}$ be the cusps of Γ, identified with vertices $x_1, x_2, \ldots, x_n$ at infinity of a Dirichlet region D for Γ. Let $g_1, g_2, \ldots, g_n \in \mathrm{SL}(2, \mathbb{R})$ be such that $x_i = g_i \cdot \infty$, $1 \leq i \leq n$. For $r > 0$, let

$$E_i(r) = \bigcup_{r < s < \infty} g_i a(s)NK.$$

Observe that $E_i(r)$ is an open subset of G. Observe also that, for $c > 1$ and $r = (\log c)/2$,

$$\{z \in \mathbf{H}^2|\ \mathrm{Im}z > c\} = \bigcup_{r < s < \infty} a(s)NK \cdot z_0,$$

where $z_0 = i$. Hence, the following proposition is an immediate consequence of Chap. II, Theorem 2.19 and Remark 2.21:

3.7 Proposition. *Let* $X = \Gamma \setminus \mathrm{SL}(2, \mathbb{R})$, *and let* $\pi : G \to X$ *be the canonical projection. For* $r_0 > 0$ *sufficiently large, the following holds:*
(i) $\pi(E_i(r_0)) \cap \pi(E_j(r_0)) = \emptyset$ *for all* $1 \le i, j \le n$ *with* $i \ne j$;
(ii) *If* $\gamma_1 E_i(r_0) \cap \gamma_2 E_i(r_0) \ne \emptyset$ *for some* $\gamma_1, \gamma_2 \in \Gamma$, *then* $\gamma_1 E_i(r_0) = \gamma_2 E_i(r_0)$.

In particular, X *decomposes as a disjoint union*

$$X = X_{r_0} \cup \bigcup_{i=1}^{n} \pi(E_i(r_0)),$$

where

$$X_{r_0} = X \setminus \bigcup_{i=1}^{n} \pi(E_i(r_0))$$

is a compact subset of X.

3.8 Corollary. *Let* r_0 *be as in the previous proposition. Let*

$$[a, b] \to G, \quad t \mapsto g_t$$

be a continuous curve in G *such that*

$$\Gamma g_t \in \bigcup_{i=1}^{n} \pi(E_i(r_0)),$$

for all $t \in [a, b]$. *Then there exist* $1 \le i \le n$ *and* $\gamma \in \Gamma$ *such that* $\gamma g_t \in E_i(r_0)$ *for all* $t \in [a, b]$.

Proof For $\gamma \in \Gamma$, the sets $\gamma E_i(r_0)$ are open and, by continuity, the set $\{g_t \mid t \in [a, b]\}$ is connected. The claim is now clear, by the previous proposition. □

3.9 Proposition. *Let* r_0 *be as in the previous proposition. If* $x \in X$ *has a non-periodic* N-*orbit, then there exists a sequence* $\{\tau_n\}_n$ *of positive real numbers, converging to infinity, such that* $xa(\tau_n)$ *lies in the compact set* X_{r_0} *for all* $n \in \mathbb{N}$.

Proof Suppose to the contrary that there exists $\tau_0 > 0$ such that

$$xa(\tau) \in \bigcup_{i=1}^{n} \pi(E_i(r_0))$$

for all $\tau \geq \tau_0$, that is, for some $g \in \pi^{-1}(\{x\})$,

$$ga(\tau) \in \bigcup_{i=1}^{n} \Gamma E_i(r_0),$$

for all $\tau \geq \tau_0$. By the previous corollary, there exist $1 \leq i \leq n$ and $\gamma \in \Gamma$ such that $ga(\tau) \in \gamma E_i(r_0)$ for all $\tau \geq \tau_0$. Set $h = g_i^{-1}\gamma^{-1}g$. So,

$$ha(\tau) \in \bigcup_{r_0 < r < \infty} a(r)NK, \qquad (8)$$

for all $\tau \geq \tau_0$. Let $(z_0, \zeta_0) \in T^1\mathbf{H}^2$ be the unit tangent vector at $z_0 = i$ pointing in the positive direction of the imaginary axis. Recall that G may be identified with the unit tangent bundle of $\mathbf{H}^2$ by means of the map $g \mapsto g \cdot (z_0, \zeta_0)$ and that $ga(\tau)$ corresponds to the image $\varphi_{2\tau}(z, \zeta)$ of $(z, \zeta) = g \cdot (z_0, \zeta_0)$ under the geodesic flow φ_τ (see Chap. II, Proposition 3.2). Hence, (8) implies that the geodesic through the point $h \cdot i \in \mathbf{H}^2$ in the direction of the vector $h \cdot \zeta_0$ eventually lies in a subset of the form

$$\{z \in \mathbf{H}^2 | \; \mathrm{Im} z > c > 0\}.$$

Of course, this is possible only if this geodesic is parallel to the imaginary axis, that is, $h \in P = AN$ (see the formula for the differential of a Möbius transformation in Chap. II, proof of Lemma 1.4). Hence, $g \in \Gamma g_i AN$. But then, by Proposition 1.6, $x = \pi(g)$ has a periodic orbit under N, a contradiction. $\square$

As a last ingredient, we shall use the following consequence of Birkhoff's ergodic theorem and of Egoroff's theorem. Recall that Egoroff's theorem asserts that if $\{f_n\}_n$ is a sequence of complex measurable functions on a space X with finite measure μ converging pointwise at every point of X, then, for every $\eta > 0$, there exists a measurable set $Y_\eta \subseteq X$ with $\mu(X - Y_\eta) < \eta$ such that $\{f_n\}_n$ converges uniformly on Y_η (see [Rd], Chap. 3, Exercise 16).

3.10 Lemma. *Let $\{t_n\}_n$ be a sequence of positive reals with $t_n \to \infty$. Let $f \in C_c(X)$. Then, for any $\eta > 0$, there exists a measurable set $Y_\eta \subseteq X$ with $\nu_G(X - Y_\eta) < \eta$ with the following property. For every $\varepsilon > 0$, there exists $n_0 \in \mathbb{N}$ such that, for all $n \geq n_0$ and for all $y \in Y_\eta$,*

$$\left| S_f(y, t_n) - \int_X f d\nu_G \right| < \varepsilon. \qquad (9)$$

Proof The horocycle flow is ergodic with respect to the G-invariant measure ν_G. The claim follows now from Birkhoff's pointwise ergodic theorem and Egoroff's theorem. $\square$

We can now give the proof of the classification theorem.

Proof of Theorem 3.1 Let $x \in X = \Gamma \setminus \mathrm{SL}(2,\mathbb{R})$ with a non-periodic N-orbit. Let $f \in C_c(X)$, viewed as a function on G. Observe that f is right uniformly continuous on G.

Fix r_0 as in Proposition 3.7. By Proposition 3.9 above, there exists a sequence $\{\tau_n\}_n$ converging to infinity such that $xa(\tau_n)$ lies in the compact subset X_{r_0} of X, for all $n \in \mathbb{N}$. Define

$$t_n := e^{2\tau_n}, \quad n \in \mathbb{N}.$$

We claim that

$$S_f(x,t_n) \to \int_X f d\nu_G, \quad \text{as} \quad n \to \infty.$$

By Proposition 3.3, this will finish the proof.

For $g \in G$ and $\delta, t > 0$, let

$$\mathbf{V}_r(g,\delta,t) := \{yu(\Theta(y,s))|\ y \in \mathbf{W}(g,\delta,t), 0 \leq s \leq r\}, \quad 0 \leq r \leq t.$$

Let $\varepsilon > 0$. Choose $\delta(\varepsilon, f)$ according to Lemma 3.6 so that inequality (7) holds. Fix $\delta > 0$ with $\delta \leq \delta(\varepsilon, f)$ such that the projection π from G to X is injective on $\mathbf{W}(g,\delta,1)$ for all $g \in \pi^{-1}(X_{r_0})$.

The mapping

$$g \mapsto \nu_G(\pi(\mathbf{V}_\varepsilon(g,\delta,1)))$$

is continuous, constant on Γ-right cosets and has strictly positive values. (Indeed, $\pi(\mathbf{V}_r(g,\delta,1))$ contains a neighbourhood of $\pi(g)$, as N^-AN is open.) Hence, since X_{r_0} is compact, there exists $\eta > 0$ such that

$$\nu_G(\pi(\mathbf{V}_\varepsilon(g,\delta,1))) > \eta \tag{10}$$

for all $g \in \pi^{-1}(X_{r_0})$. By Lemma 3.10 above, there exists $t_0 > 0$ and $Y_\eta \subseteq X$ with $\nu_G(X - Y_\eta) < \eta$ and

$$\left|S_f(y,t) - \int_X f d\nu_G\right| < \varepsilon, \tag{11}$$

for all $y \in Y_\eta$ and $t \geq t_0$.

Choose n_0 such that $t_n \geq t_0$ for all $n \geq n_0$. Set $x_n = xa(\tau_n)$, and let $g \in \pi^{-1}(x)$, $g_n \in \pi^{-1}(x_n)$ with $g_n = ga(\tau_n)$. Since (see remarks following Lemma 3.4)

$$\mathbf{W}(g,\delta,t_n) = \mathbf{W}(g_n,\delta,1)a(-\tau_n)$$

and

$$\Theta(y,s) = \Theta(y,s/t_n)t_n, \quad \forall y \in \mathbf{W}(g,\delta,t_n),$$

one has

$$\mathbf{V}_{\varepsilon t_n}(g,\delta,t_n) = \mathbf{V}_\varepsilon(g_n,\delta,1)a(-\tau_n).$$

Hence, by the G-invariance of ν_G,

$$\nu_G(\pi(\mathbf{V}_{\varepsilon t_n}(g,\delta,t_n))) > \eta, \quad \forall n \geq n_0.$$

Therefore,

$$\pi(\mathbf{V}_{\varepsilon t_n}(g,\delta,t_n)) \cap Y_\eta \neq \emptyset \qquad \forall n \geq n_0$$

and we can find $y_n \in \mathbf{W}(g,\delta,t_n)$ such that $z_n = \pi(y_n)u(\Theta(y_n,s_n)) \in Y_\eta$ for some $0 \leq s_n \leq \varepsilon t_n$. One has

$$|S_f(\pi(y_n),t_n) - S_f(z_n,t_n)| = \frac{1}{t_n}\left| \int_{-\Theta(y_n,s_n)}^{t_n-\Theta(y_n,s_n)} f(z_n u(s))ds - \int_0^{t_n} f(z_n u(s))d \right.$$

$$\leq 2\frac{\Theta(y_n,\varepsilon t_n)}{t_n}\|f\|_\infty$$

$$\leq 2\varepsilon\|f\|_\infty,$$

by the explicit formula (3) for $\Theta(y,s)$. Hence, by (7) and (11),

$$\left|S_f(\pi(y_n),\Theta(y_n,t_n)) - \int_X f d\nu_G\right| \leq |S_f(\pi(y_n),\Theta(y_n,t_n)) - S_f(\pi(y_n),t_n)| +$$

$$|S_f(\pi(y_n),t_n) - S_f(z_n,t_n)| +$$

$$\left|S_f(z_n,t_n) - \int_X f d\nu_G\right|$$

$$\leq 2\varepsilon + 2\varepsilon\|f\|_\infty + \varepsilon,$$

for $n \geq n_0$. As, by (7),

$$|S_f(\pi(y_n),\Theta(y_n,t_n)) - S_f(x,t_n)| = |S_f(y_n,\Theta(y_n,t_n)) - S_f(g,t_n)| \leq \varepsilon,$$

it follows that

$$\left|S_f(x,t_n) - \int_X f d\nu_G\right| \leq 2\varepsilon\|f\|_\infty + 4\varepsilon,$$

for all $n \geq n_0$, and this finishes the proof. $\qquad\qquad\qquad\square$

§4 Equidistribution of Horocycle Orbits

We now turn to Dani–Smillie's result [DS] on equidistribution of horocycle orbits.

Throughout this section, $G = \mathrm{SL}(2,\mathbb{R})$, and Γ is a lattice in G. Recall that

$$N = \{u(x)|\ x \in \mathbb{R}\},$$

where

$$u(x) = \begin{pmatrix} 1 & x \\ 0 & 1 \end{pmatrix}.$$

4.1 Theorem (Equidistribution of Horocycle Orbits). *Let Γ be a lattice in G. Let x be a point in $X = \Gamma \backslash G$ with a non-periodic orbit xN under the horocycle flow. Then, denoting by ν_G the unique G-invariant probability measure on X, one has*

$$\lim_{t \to \infty} \frac{1}{t} \int_0^t f(xu(s))ds = \int_X f d\nu_G$$

for every bounded continuous function f on X.

The proof is based on the classification of the ergodic measures (Theorem 3.1) and on the following theorem of independent interest.

4.2 Theorem. *Let $\varepsilon > 0$ be given. There exists a compact subset $C = C(\varepsilon)$ of $X = \Gamma \backslash G$ with the following property. If U is a unipotent one-parameter subgroup of G and if $x \in X$ has a non-periodic orbit xU, then there exists $t_0 > 0$ such that*

$$\frac{1}{t} \int_0^t \chi_C(xu(s))ds \geq 1 - \varepsilon$$

for all $t \geq t_0$, where χ_C denotes the characteristic function of C.

Observe that the theorem says, in particular, that $\{xu(t)| t > 0\}$ meets C for arbitrary large values of t. In other words, $\{xu(t)| t > 0\}$ does not tend to infinity in X as $t \to +\infty$. The result is a quantitative version for $\mathrm{SL}(2, \mathbb{R})$ of the so-called Margulis Lemma, to be discussed in the next chapter (see Chap. V, Theorem 5.1).

We first prove Theorem 4.2.

Proof of Theorem 4.2

We may of course assume that Γ is a non-uniform lattice, the theorem being obvious if Γ is uniform.

The proof depends in a crucial way on the following elementary lemma about lengths of intersections of lines with discs

$$D(0, s) = \{x = (x_1, x_2) \in \mathbb{R}^2 | \, \|x\|^2 = x_1^2 + x_2^2 < s^2\}$$

centered at 0 in the plane $\mathbb{R}^2$.

4.3 Lemma. *Let $\varepsilon > 0$ and $s_0 > 0$ be given. For $\delta = \varepsilon/(2 + \varepsilon)$, the following holds:*
Let $f : \mathbb{R} \to \mathbb{R}^2$ be a non-constant affine linear mapping with $\|f(t_0)\| = s_0$ for some $t_0 \in \mathbb{R}$ and $\|f(t)\| > s_0$ for all $t < t_0$. Then, denoting by λ the Lebesgue measure on $\mathbb{R}$, one has, for any $t > t_0$,

$$\lambda(\{\tau \in [t_0, t]| \, \|f(\tau)\| < \delta s_0\}) \leq \varepsilon \lambda(\{\tau \in [t_0, t]| \, \|f(\tau)\| < s_0\}).$$

Proof Let $\alpha, \beta \in \mathbb{R}$ with $\alpha^2 + \beta^2 \neq 0$ and $a, b \in \mathbb{R}$ be such that $f(t) = (\alpha t + a, \beta t + b)$ for all $t \in \mathbb{R}$. The line $L = \{f(t) \mid t \in \mathbb{R}\}$ intersects the boundary of the disc $D(0, s_0)$ at times $t_0 \leq t_1$. By the affine change of variable

$$t \mapsto \frac{t}{\alpha^2 + \beta^2} + t_0,$$

we may assume that $\alpha^2 + \beta^2 = 1$ and that $t_0 = 0$. So, $a^2 + b^2 = s_0^2$ and

$$\|f(t)\|^2 = t^2 + 2(\alpha a + \beta b)t + s_0^2.$$

Hence,

$$t_1 = -2(\alpha a + \beta b).$$

Set $\tau = -(\alpha a + \beta b) \geq 0$. If $\tau^2 \geq (1 - \delta^2)s_0^2$, then L intersects the boundary of the disc $D(0, \delta s_0)$ at the times

$$t^\pm = \tau \pm \sqrt{\tau^2 - (1 - \delta^2)s_0^2}.$$

One has

$$\frac{t^+ - t^-}{t^-} = \frac{2\sqrt{\tau^2 - (1 - \delta^2)s_0^2}}{\tau - \sqrt{\tau^2 - (1 - \delta^2)s_0^2}}$$

$$\leq \frac{2\sqrt{\tau^2 - (1 - \delta^2)s_0^2}(\tau + \sqrt{\tau^2 - (1 - \delta^2)s_0^2})}{(1 - \delta^2)s_0^2}.$$

Since, by the Cauchy–Schwarz inequality,

$$\tau^2 = (\alpha a + \beta b)^2 \leq (a^2 + b^2) = s_0^2,$$

it follows that

$$\frac{t^+ - t^-}{t^-} \leq \frac{2\delta(1 + \delta)}{1 - \delta^2} = \frac{2\delta}{1 - \delta} = \varepsilon.$$

Clearly, this proves the claim. □

Let $X = \Gamma \backslash \mathrm{SL}(2, \mathbb{R})$. Let $x_1, x_2, \ldots, x_n \in \mathbb{R} \cup \{\infty\}$ be the cusps of Γ, identified with vertices $x_1, x_2, \ldots, x_n$ at infinity of a Dirichlet region D for Γ. Let $g_1, g_2, \ldots, g_n \in \mathrm{SL}(2, \mathbb{R})$ be such that $x_i = g_i \cdot \infty$, $1 \leq i \leq n$. Recall from Proposition 3.7 that for $r_0 > 0$ sufficiently large, the open subsets

$$E_i(r_0) = \bigcup_{r_0 < r < \infty} g_i a(r) N K \qquad (1 \leq i, j \leq n)$$

of G have the following property: If $\gamma_1 E_i(r_0) \cap \gamma_2 E_j(r_0) \neq \emptyset$ for some $\gamma_1, \gamma_2 \in \Gamma$, then $i = j$ and $\gamma_1 E_i(r_0) = \gamma_2 E_i(r_0)$.

For $s > 0$, let

$$A_s := \{a(\tau) \mid \tau < (\log s)/2\} = \left\{ \begin{pmatrix} a & 0 \\ 0 & a^{-1} \end{pmatrix} \Big| 0 < a < \sqrt{s} \right\}.$$

For $x = \Gamma g \in X$ and $r > 0$, observe that $x \in \Gamma E_i(r)$ if and only if $g^{-1}\gamma^{-1}g_i \in KA_sN$ for some $\gamma \in \Gamma$, where $s = e^{-2r}$

4.4 Lemma. *Let $\varepsilon > 0$ be given. Let $\delta = \varepsilon/(2 + \varepsilon)$ and $r = -(\log \delta)/2$. Let r_0 be as above, and let $s_0 = e^{-2r_0}$. Let $U = \{u(t)| t \in \mathbb{R}\}$ be a unipotent subgroup of G, and let $x = \Gamma g \in X$ with a non-periodic orbit xU. Assume that $gu(t_0) \in \gamma E_i(r_0)$ for some $t_0 \in \mathbb{R}$, some $1 \le i \le n$ and some $\gamma \in \Gamma$. Then there exist $a < b$ with $a < t_0 < b$ such that the following holds:*
(i) *$gu(t) \in \gamma E_i(r_0)$ for all $t \in (a, b)$;*
(ii) *$gu(t) \notin \gamma E_i(r_0)$ for all $t \notin (a, b)$;*
(iii) *For any $a < t < b$, one has*

$$\lambda(\{\tau \in [a, t]| \, gu(\tau) \in \gamma E_i(r + r_0)\}) \le \varepsilon\lambda(\{\tau \in [a, t]| \, xu(\tau) \in \gamma E_i(r_0)\}).$$

Proof By the above remark, $gu(t) \in \gamma E_i(r_0)$ if and only if

$$u(t)^{-1}g^{-1}\gamma^{-1}g_i \in KA_{s_0}N$$

for $s_0 = e^{-2r_0}$. Similarly, $gu(t) \in \gamma E_i(r + r_0)$ if and only if

$$u(t)^{-1}g^{-1}\gamma^{-1}g_i \in KA_{\delta s_0}N.$$

Consider the natural action of $\mathrm{SL}(2, \mathbb{R})$ on $\mathbb{R}^2$. The stabilizer of $e_1 = (1, 0)^t$ under this action is N, and

$$KA_s e_1 = D(0, s) \qquad \forall s > 0.$$

Hence, for any $t \in \mathbb{R}$, one has

$$u(t)^{-1}g^{-1}\gamma^{-1}g_i \in KA_{s_0}N$$

if and only if $\|f(t)\| < s_0$, where f is the mapping

$$f : \mathbb{R} \to \mathbb{R}^2, \qquad t \mapsto (u(t)^{-1}g^{-1}\gamma^{-1}g_i) \cdot e_1.$$

As U is unipotent, f is affine linear.

We claim that f is non constant. Indeed, assume, by contradiction, that this not the case. Then

$$g_i^{-1}\gamma gu(t)^{-1}g^{-1}\gamma^{-1}g_i \in N \qquad \forall t \in \mathbb{R}. \tag{$*$}$$

The unipotent subgroup U is conjugate to N, that is, $U = h^{-1}Nh$ for some $h \in G$. By $(*)$, $g_i^{-1}\gamma gh^{-1}$ normalizes N. Hence, $g_i^{-1}\gamma gh^{-1} \in P$, where P is the group of upper triangular matrices. On the other hand, the periodic points for U in X are the points in

$$\bigcup_{1 \le j \le n} \Gamma g_j Ph$$

(see Proposition 1.6). Thus, $x = \Gamma g$ has a periodic U-orbit, a contradiction.

The claims (i) and (ii) are now clear, and (iii) follows from Lemma 4.3.
□

We are now ready for the proof of Theorem 4.2.

Proof of Theorem 4.2 Fix $\varepsilon > 0$. Let r_0 and r be as in Lemma 4.4.
Let

$$C = C(\varepsilon) = X \setminus \bigcup_{1 \le i \le n} \pi(E_i(r + r_0)).$$

Then C is a compact subset of X, and we claim that it satisfies the
requirements of Theorem 4.2.

Let $U = \{u(t) \mid t \in \mathbb{R}\}$ be a unipotent subgroup of G, and let $x = \Gamma g$
in X with a non-periodic orbit xU. For each $1 \le i \le n$, choose a sequence
$\{\gamma_k^{(i)}\}_k$ in Γ such that

$$\Gamma E_i(r_0) = \bigcup_k \gamma_k^{(i)} E_i \quad \text{and} \quad \gamma_k^{(i)} E_i \cap \gamma_l^{(i)} E_i = \emptyset \quad \forall k \ne l$$

(see Proposition 3.7). By Lemma 4.4, for each $1 \le i \le n$ and each $k \in \mathbb{N}$,
the set

$$\{s \in \mathbb{R} \mid gu(s) \in \gamma_k^{(i)} E_i(r_0)\}$$

is an open (possibly empty) interval $(a_k^{(i)}, b_k^{(i)})$ and, for any τ in $(a_k^{(i)}, b_k^{(i)})$,
one has

$$\lambda(\{s \in [a_k^{(i)}, \tau] \mid gu(s) \in \gamma_k^{(i)} E_i(r+r_0)\}) \le \varepsilon \lambda(\{s \in [a_k^{(i)}, \tau] \mid gu(s) \in \gamma_k^{(i)} E_i(r_0)\})$$

Observe that the intervals $(a_k^{(i)}, b_k^{(i)})$ are pairwise disjoint. Fix $t > 0$, and
let

$$S_0 = \{s \in [0, t] \mid xu(s) \in \bigcup_{1 \le i \le n} \pi(E_i(r_0))\} \quad \text{and}$$

$$S_1 = \{s \in [0, t] \mid xu(s) \in \bigcup_{1 \le i \le n} \pi(E_i(r + r_0))\}.$$

Let $\mathcal{F}$ be the family of all intervals $(a_k^{(i)}, b_k^{(i)})$ contained in S_0. Set $\tau_0 = b_k^{(i)}$ if 0 lies in some interval $(a_k^{(i)}, b_k^{(i)})$ and set $\tau_0 = 0$ otherwise. Let
$I_0 = (0, \tau_0)$. Further, set $I_1 = (a_k^{(i)}, t)$ if t lies in some interval $(a_k^{(i)}, b_k^{(i)})$
and set $I_1 = \emptyset$ otherwise. Then S_0 is a disjoint union of intervals

$$S_0 = I_0 \cup \bigcup_{I \in \mathcal{F}} I \cup I_1.$$

Hence, by 4.4 Lemma (iii),

$$\lambda(S_1) = \lambda(S_1 \cap I_0) + \sum_{I \in \mathcal{F}} \lambda(S_1 \cap I) + \lambda(S_1 \cap I_1)$$

$$\leq \tau_0 + \varepsilon \sum_{I \in \mathcal{F}} \lambda(I) + \varepsilon \lambda(I_1)$$

$$= \tau_0 + \varepsilon \lambda(S_0 \cap [\tau_0, t]).$$

As $\lambda(S_0 \cap [\tau_0, t]) \leq t - \tau_0$, it follows that

$$\frac{\lambda(S_1)}{t} \leq \frac{(1 - \varepsilon)\tau_0}{t} + \varepsilon.$$

Hence, for $t_0 > 0$ large enough, one has

$$\frac{\lambda(S_1)}{t} \leq 2\varepsilon,$$

for all $t > t_0$, that is,

$$\frac{1}{t} \int_0^t \chi_C(xu(s)) ds \geq 1 - 2\varepsilon.$$

This completes the proof. $\qquad\qquad\qquad\qquad\qquad\qquad\qquad\square$

Proof of the Equidistribution Theorem

Let $C_0(X)$, the space of the continuous functions vanishing at infinity on $X = \Gamma \setminus G$, be equipped with the supremum norm. Recall that its dual space $C_0^*(X)$ may be identified with the Borel measures on X (Riesz representation theorem, see Chap. I, §3). For $x \in X$, let $T_{x,t}$ be the linear form on $C_0(X)$ defined by

$$T_{x,t} f = \frac{1}{t} \int_0^t f(xu(s)) ds, \qquad f \in C_0(X).$$

$T_{x,t}$ is a positive operator with $\|T_{x,t}\| = 1$.

Let $\mu \in C_0(X)^*$ be a weak-$*$-limit point of the set $\{T_{x,t} | t > 0\}$. Then, clearly μ is N-invariant, positive and $\|\mu\| = \mu(X) \leq 1$. In fact, one has the following consequence of Theorem 4.2.

4.5 Corollary. *Let x be a point in X with a non-perodic orbit xN. Let $\mu \in C_0(X)^*$ be such that $T_{x,t_n} \to \mu$ in the weak-$*$-topology for a sequence $\{t_n\}_n \subseteq \mathbb{R}^+$ with $\lim_{n \to \infty} t_n = \infty$. Then μ is a probability measure, that is, $\|\mu\| = \mu(X) = 1$. Moreover, $T_{x,t_n}(\varphi) \to \mu(\varphi)$ for all continuous bounded functions φ on X.*

Proof Fix $\varepsilon > 0$. Let $C = C(\varepsilon)$ be a compact subset of X as in Theorem 4.2. Let φ be a continuous function on X with compact support such that

$$\chi_C \leq \varphi \leq 1.$$

Then, for $n \in \mathbb{N}$ large enough,

$$1 - \varepsilon \leq T_{x,t_n}(\chi_C) \leq T_{x,t_n}(\varphi).$$

Since $T_{x,t_n}(\varphi) \to \mu(\varphi)$, this implies that

$$1 - \varepsilon \leq \mu(\varphi).$$

Therefore $1 - \varepsilon \leq \mu(X)$. Hence, $\mu(X) = 1$.

The last statement is a well-known consequence of the fact that μ is a probability measure. Indeed, let φ be a continuous bounded function on X, and fix $\varepsilon > 0$. There exists a continuous function ψ on X with compact support such that

$$0 \leq \psi \leq 1 \quad \text{and} \quad 1 - \mu(\psi) \leq \varepsilon/2.$$

As $T_{x,t_n}(\psi) \to \mu(\psi)$, one has $1 - T_{x,t_n}(\psi) \leq \varepsilon$ and, hence,

$$|T_{x,t_n}(\varphi(1 - \psi))| \leq \varepsilon\|\varphi\|$$

for n large enough. Therefore, by the triangle inequality,

$$|T_{x,t_n}(\varphi) - \mu(\varphi)| \leq 2\varepsilon\|\varphi\| + |T_{x,t_n}(\varphi\psi) - \mu(\varphi\psi)|,$$

for n large enough. As $\varphi\psi$ has compact support,

$$T_{x,t_n}(\varphi\psi) \to \mu(\varphi\psi).$$

Hence, $T_{x,t_n}(\varphi) \to \mu(\varphi)$ and this proves the claim. $\square$

We shall need the following general criterion about convergence to infinity of sequences in $X = \Gamma \backslash G$ (see [Rag], Chap.I, Theorem 1.12).

4.6 Lemma. *Let G be a second countable locally compact group, and let Γ be a lattice in G. Let $\pi : G \to \Gamma \backslash G$ be the canonical projection and $X = \Gamma \backslash G$. For a sequence $\{x_n = \pi(g_n)\}_{n \in \mathbb{N}} \in X$, the following are equivalent:*

(i) *$\{x_n\}_n$ has no convergent subsequence;*

(ii) *There exists a sequence $\{\gamma_n\}_{n \in \mathbb{N}} \in \Gamma$, $\gamma_n \neq e$, such that*

$$\lim_n g_n^{-1}\gamma_n g_n = e.$$

Proof We first show that (i) implies (ii). Let ν denote the Haar measure on G as well as the corresponding invariant measure on $X = \Gamma \backslash G$. Let $\{K_n\}_n$ be a sequence of compact subsets of G with

$$G = \bigcup_{n \in \mathbb{N}} K_n.$$

Set

$$\varepsilon_n = \nu(X \backslash \pi(K_n)).$$

Then $\lim_{n\to\infty} \varepsilon_n = 0$. Let V_n be a fundamental system of compact neighbourhoods of the group unit e in G with $\nu(V_n) > \varepsilon_n$ for all $n \in \mathbb{N}$. Then $W_n = V_n V_n^{-1}$ is also a fundamental system of compact neighbourhoods of e.

Now, $\pi(K_n W_n)$ is compact and the sequence $\{x_n\}_n$ has no limit points. Hence, there exists for each n a positive integer i_n with

$$x_i \notin \pi(K_n W_n), \qquad \forall i \geq i_n.$$

This clearly implies that

$$x_i V_n \cap \pi(K_n) V_n = \emptyset, \qquad \forall i \geq i_n.$$

Thus,

$$\nu(x_i V_n) \leq \nu(X \setminus \pi(K_n) V_n) \leq \nu(X \setminus \pi(K_n)) \leq \varepsilon_n, \qquad \forall i \geq i_n.$$

As, on the other hand, $\nu(g_i V_n) = \nu(V_n) > \varepsilon_n$, we can find for each $i \geq i_n$ elements $\gamma_i \in \Gamma$, $\gamma_i \neq e$, and $v, v' \in V_n$ such that $\gamma_i g_i v = g_i v'$. Hence,

$$g_i^{-1} \gamma_i g_i \in W_n, \qquad \forall i \geq i_n.$$

As W_n is a fundamental system of neighbourhoods of e, we may clearly choose $\gamma_i \in \Gamma$, $\gamma_i \neq e$ such that $g_i^{-1} \gamma_i g_i \to e$ as $i \to \infty$.

To show that (ii) implies (i), let $\gamma_n \in \Gamma$, $\gamma_n \neq e$ with $g_n^{-1} \gamma_n g_n \to e$ as $n \to \infty$. Assume, by contradiction, that $\{x_n = \pi(g_n)\}_{n\in\mathbb{N}}$ has a limit point $x = \pi(g) \in X$. Upon passing to a subsequence, we may assume that $\theta_n g_n \to g$ for some sequence $\theta_n \in \Gamma$. Set $w_n = g_n^{-1} \gamma_n g_n$, and write

$$w_n = g_n^{-1} \theta_n^{-1} \theta_n \gamma_n \theta_n^{-1} \theta_n g_n.$$

As $w_n \to e$ and as $\theta_n g_n \to g$, we see that $\theta_n \gamma_n \theta_n^{-1} \to e$. Since Γ is discrete and since $\theta_n \gamma_n \theta_n^{-1} \in \Gamma$, it follows that eventually $\gamma_n = e$. This is a contradiction. $\square$

4.7 Proposition. *Let D be a compact subset of X containing only points with a periodic N-orbit. Let C be another compact subset of X. There exists τ_0 such that*

$$Da(\tau) \subseteq X \setminus C, \qquad \forall \tau \geq \tau_0.$$

Proof Let $x = \pi(g) \in X$ be a point with a periodic orbit xN. Then, for some $t_0 > 0$ and $\gamma \in \Gamma$, $gu(t_0) = \gamma g$. Thus $g^{-1} \gamma g = u(t_0)$. Observe that $\gamma \neq e$. Now,

$$(ga(\tau))^{-1} \gamma ga(\tau) = a(\tau)^{-1} u(t_0) a(\tau) = u(e^{-2\tau} t_0) \to e$$

as $\tau \to +\infty$. Then the previous lemma shows that $xa(\tau)$ eventually leaves any compact subset C of X. The claim follows now from an obvious compactness argument. $\square$

Now we are ready to give the proof of the equidistribution theorem.

Proof of Theorem 4.1   Fix $x \in X = \Gamma \backslash G$ with a non-periodic orbit xN. Let ν_G denote the G–invariant probability measure on X. We have to show that, in the weak–$*$–topology,

$$T_{x,t} \to \nu_G, \qquad \text{as } t \to +\infty,$$

For this, it clearly suffices to prove that ν_G is the only weak–$*$–limit point of the set $\{T_{x,t} | t > 0\}$ as $t \to \infty$.

Let μ be such a limit point. Let $\{t_n\}_n \subseteq \mathbb{R}^+$ be a sequence with $\lim_{n \to \infty} t_n = \infty$ such that $T_{x,t_n} \to \mu$ in the weak–$*$–topology.

Clearly, μ is N-invariant. By Corollary 4.5, we know that μ is a probability measure on X. By the classification of ergodic N-invariant measures (Theorem 3.1), it remains to show that $\mu(Y) = 0$, where Y is the union of all periodic N-orbits.

Suppose, by contradiction, that $\beta = \mu(Y) > 0$. Let $0 < \varepsilon < \beta/4$, and let $C = C(\varepsilon)$ be as in Theorem 4.2. Let D be a compact subset of Y with $\mu(D) \geq 3\beta/4$. As $T_{x,t_n} \to \mu$, one has

$$\lim_{n \to \infty} T_{x,t_n}(\varphi) = \mu(\varphi) \geq \mu(D) \geq \frac{3\beta}{4}$$

for all continuous bounded functions φ on X with $\varphi \geq \chi_D$ (see Corollary 4.5). By the previous proposition, there exists $\tau > 0$ such that

$$Da(\tau) \subseteq X \backslash C.$$

Set $z = xa(\tau)$. Then

$$\lim_{n \to \infty} T_{z,t_n e^{-2\tau}}(\varphi) = \lim_{n \to \infty} T_{x,t_n}(\varphi_{a(\tau)}) \geq \frac{3\beta}{4},$$

for all continuous bounded functions φ on X with $\varphi \geq \chi_{Da(\tau)}$, where $\varphi_{a(\tau)}$ is the function defined by $\varphi_{a(\tau)}(y) = \varphi(ya(\tau))$ for all $y \in X$. Hence,

$$\lim_{n \to \infty} T_{z,t_n e^{-2\tau}}(\psi) \leq 1 - \frac{3\beta}{4}, \tag{$*$}$$

for all continuous bounded functions ψ on X with $\psi \leq \chi_C$, since

$$1 - \psi \geq \chi_{Da(\tau)}.$$

Choose a continuous bounded function ψ on X with

$$0 \leq \chi_C - \psi \leq \frac{\beta}{4}.$$

Then, for all $t > 0$,

$$T_{z,t}(\chi_C) - T_{z,t}(\psi) \leq \frac{\beta}{4}. \tag{$**$}$$

By $(*)$, there exists $n_0 \in \mathbb{N}$ such that

$$T_{z,t_n e^{-2\tau}}(\psi) \leq 1 - \frac{\beta}{2}, \qquad (***)$$

for all $n \geq n_0$. It follows from $(**)$ and $(***)$ that

$$T_{z,t_n e^{-2\tau}}(\chi_C) \leq 1 - \frac{\beta}{4} < 1 - \varepsilon,$$

for $n \geq n_0$. But then, Theorem 4.2 forces zN to be a periodic orbit. Hence, so is $xN = zNa(\tau)^{-1}$. This contradiction finishes the proof of the theorem. □

Notes

For a survey on the differences in behaviour of the geodesic and the horocyclic flow, see [Ra1]. The horocyclic flow has been considered by E. Hopf ([Ho1], [Ho2]) during his study of the geodesic flow. Ergodicity and mixing of the horocyclic flow were proved by G. Hedlund [He1], as well as its minimality.

The simple but useful Lemma 2.2 about minimal invariant sets appears in [Ma3].

The result in Exercise 2.7 was proved by K. Mahler [Mah2] under the additional assumption that the bilinear form f is positive definite. For a further generalization, see [AGH], Chap. IX, Theorem 1.

The unique ergodicity of the horocyclic flow (in the cocompact case) was first proved by H. Furstenberg [Fu1]. Generalizations of this result appear in [Bow], [EP] and [Ve2]. The classification of ergodic measures in the non-cocompact case was proved by [Da1]. The uniform distribution result was obtained by S. G. Dani and J. Smillie [DS]. The (completely different) proofs we give for the classification of ergodic measures and for the equidistribution theorem follow [Ra7]. In fact, all the above results have been generalized by M. Ratner in her spectacular work on the so called Raghunathan conjectures, to be discussed in Chapter VI.

Theorem 4.2 is a special case of much more general results established by Dani ([Da2], [Da3],[Da7]; see the notes at the end of Chapter V). The short and elementary proof we gave for Theorem 4.2 was suggested to us by the referee.

Consider a surface $\Sigma = \Gamma \backslash \mathbf{H}^2$, where Γ a finitely generated discrete group of $\mathrm{PSL}(2,\mathbb{R})$ acting freely on $\mathbf{H}^2$. By Ratner's work (see Chap. VI, Theorem 6.1), the classification result of the probability measures on $T^1(\Sigma)$ which are invariant under the horocycle flow (Theorem 3.1) is valid even if $\mathrm{area}(\Sigma) = \infty$. In this case, the only invariant finite ergodic measures are supported on periodic orbits. In particular, if Σ has no cusps, there is no invariant probability measure on $T^1(\Sigma)$. Concerning *infinite* measures on

$T^1(\Sigma)$ which are invariant under the horocycle flow, the situation is not so well understood. The complete classification of such measures has been given by M. Burger [Bu] in the case where Σ is geometrically finite and has no cusps. His approach is based on the theory of unitary representations of $\text{PSL}(2,\mathbb{R})$. This allows him to prove, for a compact surface Σ, an equidistribution theorem similar to Theorem 4.1, but valid uniformly for all $x \in \Gamma \setminus \text{SL}(2,\mathbb{R})$ and with an estimate for the error term involving the first non-zero eigenvalue of the Laplacian on Σ (see [Bu], Theorem 2). In particular, this gives a purely representation theoretic proof of the unique ergodicity result of Furstenberg (see Corollary 3.2).

M. Babillot and F. Ledrappier [BL] construct an infinite family of ergodic measures on $T^1(\Sigma')$ which are invariant under the horocycle flow for the homology cover $\Sigma' = [\Gamma, \Gamma] \setminus \mathbf{H}^2$ of a compact surface $\Sigma = \Gamma \setminus \mathbf{H}^2$.

Chapter V

Siegel Sets, Mahler's Criterion and Margulis' Lemma

In this chapter, we shall be concerned with the homogeneous space $\mathrm{SL}(n,\mathbb{R})/\mathrm{SL}(n,\mathbb{Z})$ which may be realized as the space of all unimodular lattices in $\mathbb{R}^n$ (see remarks before Exercise 3.1 below).

In Section 1, we introduce Siegel sets and show that they are approximations to fundamental domains for the action of $\mathrm{SL}(n,\mathbb{Z})$ by translations on $\mathrm{SL}(n,\mathbb{R})$. As shown in Section 2, this gives a proof that $\mathrm{SL}(n,\mathbb{Z})$ is a lattice in $\mathrm{SL}(n,\mathbb{R})$. As another application, Siegel sets are used in Section 3 in order to prove Mahler's criterion for compactness of sets of lattices in $\mathbb{R}^n$. In Section 4, Siegel sets are applied to the reduction theory of positive definite quadratic forms.

Section 5 is devoted to Margulis' Lemma which – at least, in its original version – deals with recurrence properties of orbits of unipotent subgroups of $\mathrm{SL}(n,\mathbb{R})$ in the homogeneous space $\mathrm{SL}(n,\mathbb{R})/\mathrm{SL}(n,\mathbb{Z})$. It says that an orbit under such a subgroup never tends to infinity in $\mathrm{SL}(n,\mathbb{R})/\mathrm{SL}(n,\mathbb{Z})$. Via Mahler's criterion, this fact may be reformulated in terms of orbits of lattices in $\mathbb{R}^n$.

§1 Siegel Sets in $\mathrm{SL}(n,\mathbb{R})$

Let $G = \mathrm{SL}(n,\mathbb{R})$ and let $\Gamma = \mathrm{SL}(n,\mathbb{Z})$. This notation will be adopted throughout this section. We shall consider the action of Γ by *right* translations on G.

1.1 Exercise. Show that Γ is a discrete subgroup of G.

Let $K = \mathrm{SO}(n,\mathbb{R})$, let

$$A = \{\mathrm{diag}(a_1,\ldots,a_n)|\ a_1\cdots a_n = 1,\ a_i > 0,\ \forall i = 1,\ldots,n\}$$

be the subgroup of the diagonal matrices in $\mathrm{SL}(n,\mathbb{R})$ with positive diagonal entries and

$$N = \{(n_{ij})_{i,j} \in \mathrm{SL}(n,\mathbb{R})|\ n_{ii} = 1,\ n_{ij} = 0 \quad \text{for} \quad i < j\}$$

the subgroup of the upper triangular unipotent matrices.

The following elementary lemma is the Iwasawa decomposition of $G = \mathrm{SL}(n,\mathbb{R})$. (For $n = 2$, see the beginning of the previous chapter.)

1.2 Theorem. *The product map*

$$K \times A \times N \to G, \qquad (k, a, n) \mapsto kan$$

is a homeomorphism.

Proof An inverse map is defined by the usual Gram–Schmidt orthonormalization procedure as follows.

Let $x_1, x_2, \ldots, x_n$ be the columns of a matrix $g \in \mathrm{SL}(n, \mathbb{R})$ and define inductively $y_1, y_2, \ldots, y_n$ by

$$y_1 = \frac{x_1}{\|x_1\|},$$

$$\tilde{y}_k = x_k - \sum_{\ell=1}^{k-1} < x_k, y_\ell > y_\ell, \quad y_k = \frac{\tilde{y}_k}{\|\tilde{y}_k\|} \qquad (2 \geq k \geq n).$$

Let $e_1, \ldots, e_n$ be the standard basis of $\mathbb{R}^n$. Let k^{-1} be the matrix with

$$k^{-1}(y_i) = e_i, \quad 1 \leq i \leq n.$$

Then k is orthogonal, and it is readily verified that $k^{-1}g$ is upper triangular with positive entries on the diagonal.

Clearly, this defines a continuous inverse of the product map

$$(k, a, n) \mapsto kan.$$

$\square$

1.3 Definition. A *Siegel set* in $\mathrm{SL}(n, \mathbb{R})$ is a set $\Sigma_{t,u}$ of the form

$$\Sigma_{t,u} = KA_t N_u,$$

where t, u are positive real numbers,

$$A_t = \{a \in A \mid a_i/a_{i+1} \leq t, \quad i = 1, \ldots, n\},$$

and

$$N_u = \{n \in N \mid |n_{ij}| \leq u, \quad 1 \leq i < j \leq n\}.$$

Clearly, N_u is compact, whereas A_t is not. We are going to show that

$$G = \Sigma_{t,u}\Gamma$$

whenever $t \geq 2/\sqrt{3}$ and $u \geq 1/2$. It will be helpful to look first at the case $n = 2$.

1.4 Example. Let $n = 2$. So, $G = \mathrm{SL}(2, \mathbb{R})$ and $\Gamma = \mathrm{SL}(2, \mathbb{Z})$. It was shown in Chap. II, Example 2.7, that

$$D = \{z \in \mathbf{H}^2 \mid |z| \geq 1, |\mathrm{Re}z| \leq 1/2\}$$

is a fundamental domain for the action of Γ on $\mathbf{H}^2 \cong G/K$. In particular, every Γ-orbit in $\mathbf{H}^2$ meets the set

$$S = \{z \in \mathbf{H}^2|\ \mathrm{Im}z \geq \sqrt{3}/2,\ |\mathrm{Re}z| \leq 1/2\}.$$

Image of a Siegel set in the Poincaré half plane

Now $S = N_{1/2}A_{\sqrt{3}/2}i$. Hence, with $p : G \to \mathbf{H}^2$, $\quad g \mapsto g \cdot i$,

$$G = \Gamma p^{-1}(S) = \Gamma N_{1/2}A_{\sqrt{3}/2}K.$$

Taking inverses shows that

$$G = KA_{2/\sqrt{3}}N_{1/2}\Gamma = \Sigma_{2/\sqrt{3},1/2}\Gamma,$$

as claimed. Observe that translates of (the interior of) $\Sigma_{2/\sqrt{3},1/2}$ under elements from G are not necessarily disjoint. Indeed, the same is true for translates of S under the action of G on $\mathbf{H}^2$. Hence, $\Sigma_{2/\sqrt{3},1/2}$ is not a fundamental domain for the action of $\mathrm{SL}(2,\mathbb{Z})$ on $\mathrm{SL}(2,\mathbb{R})$. Observe also that the constants $2/\sqrt{3}$ and $1/2$ are the best possible.

As the following lemma shows, treating the condition on the N-part is easy. The point is the condition on the A-part. This will be done later by means of a minimum principle.

1.5 Lemma. *For* $N_\mathbb{Z} = N \cap \Gamma$, *one has* $N = N_\mathbb{Z}N_{1/2}$.

Proof This is proved by induction on n. Let

$$u = (u_{ij})_{1\leq i,j\leq n} \in N.$$

One has to show that $|(u\gamma)_{ij}| \leq 1/2$ for some $\gamma \in N_\mathbb{Z}$. Write

$$u = \begin{pmatrix} 1 & x \\ 0 & u' \end{pmatrix} \in N,$$

where u' is an $(n-1)\times(n-1)$ unipotent matrix and $x = (u_{1i})_{2\leq i\leq n}$. By induction, we can find a unipotent integer $(n-1)\times(n-1)$ matrix γ' so that $|(u'\gamma')_{ij}| \le 1/2$. Then

$$v = u \begin{pmatrix} 1 & 0 \\ 0 & \gamma' \end{pmatrix} = \begin{pmatrix} 1 & x\gamma' \\ 0 & u'\gamma' \end{pmatrix}.$$

Let $y = (y_{1i})_{2 \le i \le n} \in \mathbb{Z}^{n-1}$ be such that $|v_i + y_{1i}| \le 1/2$, where $x\gamma' = (v_i)_{2 \le i \le n}$. Then

$$\gamma = \begin{pmatrix} 1 & 0 \\ 0 & \gamma' \end{pmatrix} \begin{pmatrix} 1 & y \\ 0 & I \end{pmatrix}$$

has the required property.   □

If $g \in G$, we write $g = k_g a_g n_g$ for the Iwasawa decomposition of g. For $v = (v_1, \cdots, v_n) \in \mathbb{R}^n$, let

$$\|v\| = \sqrt{v_1^2 + \ldots + v_n^2}$$

denote the Euclidean norm of v.

1.6 Lemma. *Let* $g \in G$ *with Iwasawa decomposition* $g = kan$. *Assume that* $\|ge_1\| \le \|gv\|$ *for all* $v \in \mathbb{Z}^n \setminus \{0\}$. *Then* $a_1/a_2 \le 2/\sqrt{3}$, *where* a_i *denotes the* i*-th diagonal element of* a.

Proof Observe that $a_{gu} = a_g$ and that $\|gue_1\| = \|ge_1\|$ for any $u \in N$. So, we may assume, in view of the previous lemma, that $|n_{ij}| \le 1/2$ for all $i < j$. Notice that

$$\|ge_1\| = \|kan(e_1)\| = \|an(e_1)\| = \|ae_1\| = a_1.$$

As $ne_2 = e_2 + n_{12}e_1$, this implies that

$$a_1^2 = \|g(e_1)\|^2 \le \|a(e_2 + n_{12}e_1)\|^2 = a_1^2 n_{12}^2 + a_2^2 \le a_1^2/4 + a_2^2.$$

This proves the lemma.   □

1.7 Theorem. *For* $t \ge 2/\sqrt{3}$ *and* $u \ge 1/2$, *one has* $G = \Sigma_{t,u}\Gamma$.

Proof The proof is by induction on n. For $n = 1$ there is nothing to prove as $G = \{1\}$. Let $n > 1$, and assume that the assertion is true for $\mathrm{SL}(n-1, \mathbb{R})$. Let $g \in G$. The function

$$\mathbb{R}^n \to \mathbb{R}^+, \quad v \mapsto \|gv\|$$

takes its minimum m on $\mathbb{Z}^n \setminus \{0\}$. Indeed, $g\mathbb{Z}^n$ is a discrete subset (in fact, a lattice) of $\mathbb{R}^n$. Let $v_0 \in \mathbb{Z}^n \setminus \{0\}$ be such that

$$\|gv_0\| = \min\{\|gv\| \,|\, v \in \mathbb{Z}^n \setminus \{0\}\}.$$

If $v_0 = \alpha v$ for some $\alpha \in \mathbb{Z}, v \in \mathbb{Z}^n$, then $\alpha = 1$ or -1. In other words, v_0 is a primitive vector. As $\Gamma = \mathrm{SL}(n, \mathbb{Z})$ acts transitively on the set of such vectors, we can find $\gamma \in \Gamma$ such that $\gamma e_1 = v_0$.

Set $g' = g\gamma$. Then

$$\|g'e_1\| \leq \|g'v\|, \qquad \forall v \in \mathbb{Z}^n \setminus \{0\}.$$

It suffices to find $\gamma' \in \Gamma$ such that $g'\gamma' \in \Sigma_{t,u}$.

Let $g' = k'a'n'$ be the Iwasawa decomposition of g' . Let $h = a'n'$.
Clearly, it suffices to show that $h\gamma' \in \Sigma_{t,u}$ for some $\gamma' \in \Gamma$. In view of
Lemma 1.5 above, it suffices to find $\gamma' \in \Gamma$ such that $h\gamma' \in KA_tN$.

Now, h is of the form

$$h = \begin{pmatrix} \lambda & * \\ 0 & A \end{pmatrix}$$

with $\lambda \in \mathbb{R}^+$ and A is a triangular matrix in GL$(n-1, \mathbb{R})$. Let $\beta \in \mathbb{R}^+$
be such that

$$\beta^{n-1} = 1/\lambda = \det A.$$

Then $A = \beta A'$ for $A' \in$ SL$(n-1, \mathbb{R})$. By induction hypothesis, there
exists $\gamma'' \in$ SL$(n-1, \mathbb{Z})$ such that $A'\gamma''$ lies in $\Sigma_{t,u}^{n-1}$, the Siegel set for
SL$(n-1, \mathbb{R})$. Now set

$$\gamma' = \begin{pmatrix} 1 & 0 \\ 0 & \gamma'' \end{pmatrix}.$$

Then $\gamma' \in$ SL$(n, \mathbb{Z})$ and

$$h\gamma' = \begin{pmatrix} \lambda & * \\ 0 & \beta A'\gamma'' \end{pmatrix}.$$

Let $A'\gamma'' = k''a''n''$ be the Iwasawa decomposition of $A'\gamma''$. Then

$$kan = \begin{pmatrix} 1 & 0 \\ 0 & k' \end{pmatrix} \begin{pmatrix} \lambda & 0 \\ 0 & \beta a' \end{pmatrix} \begin{pmatrix} 1 & * \\ 0 & n' \end{pmatrix}$$

is the Iwasawa decomposition of $h\gamma'$. So, a has diagonal entries

$$a_1 = \lambda, \quad a_i = \beta a_i' \ \ (2 \leq i \leq n), \quad a_i'/a_{i+1}' \leq 2/\sqrt{3} \ \ (2 \leq i < n).$$

It remains to show that

$$a_1/a_2 \leq 2/\sqrt{3}.$$

The matrix γ' fixes e_1 . Hence,

$$\|h\gamma'(e_1)\| = \|he_1\| = \|a'n'(e_1)\| = \|k'a'n'(e_1)\| = \|g'e_1\|.$$

Since

$$\|g'e_1\| \leq \|g'v\| = \|hv\|$$

for all $v \in \mathbb{Z}^n \setminus \{0\}$, one has

$$\|h\gamma'(e_1)\| \leq \|h\gamma'(v)\|, \qquad \forall v \in \mathbb{Z}^n \setminus \{0\}.$$

The claim follows now from the previous lemma.   □

§2 SL($n, \mathbb{Z}$) **is a Lattice in** SL($n, \mathbb{R}$)

Our first application of the existence of Siegel sets will be a proof that SL($n, \mathbb{Z}$) is a lattice in SL($n, \mathbb{R}$). As above, let A be the subgroup of SL($n, \mathbb{R}$) consisting of the diagonal matrices with positive diagonal entries, and let N be the subgroup of the upper unipotent matrices.

Let B the subgroup of the upper triangular matrices with positive diagonal entries. Then B is the semi-direct product $B = AN$ with A acting on N by conjugation. We first need some elementary facts about Haar measures on these subgroups. Let da denote a Haar measure on A and dn a left Haar measure on N.

2.1 Lemma. *The measure $db = dadn$ is a left Haar measure on B.*

The proof is straightforward and left to the reader.

2.2 Lemma. *The left Haar measure dn on N may be identified with the Lebesgue measure on $\mathbb{R}^{n(n-1)/2}$ by means of the homeomorphism $n \to (n_{ij})_{1 \leq i < j \leq n}$. Moreover, dn is right invariant.*

Proof For $u \in N$, let

$$L_u : \mathbb{R}^{n(n-1)/2} \to \mathbb{R}^{n(n-1)/2}, \ x \mapsto ux$$

denote the left translation by u. The Jacobian of this transformation is easily seen to be 1. Indeed, one has for all $i < j$

$$(L_u x)_{ij} = u_{ij} + x_{ij} + \sum_{i<k<j} u_{ik} x_{kj}.$$

Endow the set of all pairs (i, j) with $1 \leq i < j \leq n$ with the lexicographical order. It is clear that the Jacobi matrix of L_u is lower triangular with diagonal entries all equal to 1.
The same is true for right translations.   □

2.3 Lemma. *The isomorphism*

$$A \to \mathbb{R}^{n-1}, \ \text{diag}(a_1, \ldots, a_n) \to (\log \frac{a_1}{a_2}, \log \frac{a_2}{a_3}, \ldots, \log \frac{a_{n-1}}{a_n})$$

identifies da with the Lebesgue measure on $\mathbb{R}^{n-1}$.

The proof is clear.

2.4 Lemma. *Let*

$$\rho(a) = \prod_{i<j} \frac{a_i}{a_j}, \qquad a = \text{diag}(a_1, \ldots, a_n) \in A.$$

Then $d_r b = \rho(a)dadn$ is a right Haar measure on B.

Proof Let $a \in A$. For $n \in N$, one has

$$(ana^{-1})_{ij} = \frac{a_i}{a_j} n_{ij}, \qquad i < j.$$

Hence, when viewed as a transformation of $\mathbb{R}^{n(n-1)/2}$, the automorphism

$$\mathrm{Ad}(a): \ n \mapsto ana^{-1}$$

has the Jacobian

$$\det \mathrm{Ad}(a) = \prod_{i<j} \frac{a_i}{a_j} = \rho(a).$$

This shows immediately that $\rho(a)dadn$ is right invariant. Indeed, for any continuous function f on B with compact support and for $a_0 n_0 \in B$,

$$\int_A \int_N f(ana_0n_0)\rho(a)dadn = \int_A \int_N f(aa_0(\mathrm{Ad}(a_0^{-1})n)n_0)\rho(a)dadn$$

$$= \int_A \int_N f(a(\mathrm{Ad}(a_0^{-1})n)n_0)\rho(a_0^{-1}a)dnda$$

$$= \int_A \int_N f(ann_0)(\rho(a_0^{-1}))^{-1}\rho(a_0^{-1}a)dnda$$

$$= \int_A \int_N f(an)\rho(a)dadn.$$

$\square$

2.5 Lemma. *Identify* $G = \mathrm{SL}(n,\mathbb{R})$ *with* $KAN = KB$, *by means of the Iwasawa decomposition, and let* dk *be a Haar measure on* K. *Then* $dkd_rb = \rho(a)dkdadn$ *is a left (and right) Haar measure on* G.

Proof Let dg be a left Haar measure on G. As G is unimodular, dg is also right invariant. Let dx be the inverse image of dg under the homeomorphism

$$K \times B \to G, \qquad (k,b) \mapsto kb.$$

The measure dx on the direct product $K \times B$ is clearly left invariant under K and right invariant under B. So, up to a scalar, dx must be the product measure dkd_rb. $\square$

2.6 Lemma. *Every Siegel set* $\Sigma_{t,u}$ *in* $\mathrm{SL}(n,\mathbb{R})$ *has finite measure with respect to a Haar measure.*

Proof By the previous lemma,

$$\int_{\Sigma_{t,u}} dg = \int_K \int_{A_t} \int_{N_u} \rho(a)dkdadn.$$

As K and N_u are compact, it suffices to show that

$$\int_{A_t} \rho(a)da < \infty.$$

Now, if we set $b_i = a_i/a_{i+1}$, then

$$\rho(a) = \prod_{i=1}^{n-1} b_i^{r_i}$$

for some positive integers r_i. So, identifying A with $\mathbb{R}^{n-1}$ as in Lemma 2.3, ρ corresponds to the function

$$(y_1, \ldots, y_n) \rightarrow \prod_{i=1}^{n-1} \exp(r_i y_i)$$

on $\mathbb{R}^{n-1}$. Hence, we have,

$$\int_{A_t} \rho(a)da = \prod_{i=1}^{n-1} \int_{-\infty}^{t} \exp(r_i y_i)dy_i.$$

This is a finite integral as $r_i > 0$.   □

2.7 Theorem. *The discrete subgroup* $\mathrm{SL}(n, \mathbb{Z})$ *is a lattice in* $\mathrm{SL}(n, \mathbb{R})$.

Proof   This follows from the previous lemma and Theorem 1.7.   □

§3 Mahler's Criterion

The second application is Mahler's criterion which will play an important rôle in the next chapter.

Let $\mathcal{L}$ be the set of *unimodular* lattices in $\mathbb{R}^n$. These are the lattices of covolume 1 in $\mathbb{R}^n$. The group $G = \mathrm{SL}(n, \mathbb{R})$ acts transitively on $\mathcal{L}$. Since $\Gamma = \mathrm{SL}(n, \mathbb{Z})$ stabilizes the standard lattice $L_0 = \mathbb{Z}^n$, the set $\mathcal{L}$ may be identified with the homogeneous space G/Γ by means of the mapping

$$\Phi : G/\Gamma \rightarrow \mathcal{L}, \quad g\Gamma \mapsto gL_0.$$

We endow $\mathcal{L}$ with the natural locally compact topology of G/Γ.

3.1 Exercise.   Show that the topology on $\mathcal{L}$ may also be described as follows: a sequence of lattices $\{L_i\}_i$ converges to a lattice L if each L_i has a basis $\{b_1^{(i)}, \ldots, b_n^{(i)}\}$ and L has a basis $\{b_1, \ldots, b_n\}$ such that

$$\lim_i b_1^{(i)} = b_1, \ldots, \lim_i b_n^{(i)} = b_n.$$

Hint: Show that the above defines a topology on $\mathcal{L}$ for which $\mathcal{L}$ is locally compact and separable. Moreover, the mapping Φ is continuous, when $\mathcal{L}$

is endowed with this new topology. It is a standard fact that this implies that Φ is a homeomorphism.

3.2 Theorem (Mahler's Compactness Criterion). *Let M be a subset of $\mathcal{L}$. The following properties are equivalent:*
(i) *M is relatively compact;*
(ii) *There exists a neighbourhood U of the origin in $\mathbb{R}^n$ such that $L \cap U = \{0\}$ for all $L \in M$.*

Proof The proof that (i) implies (ii) is elementary. Indeed, assume that M is relatively compact and suppose that there exists a sequence $\{L_i = g_i L_0\}_{i \in \mathbb{N}}$ of lattices in M such that $\lim_i x_i = 0$ for some $x_i \in L_i \setminus \{0\}$. Write $x_i = g_i y_i$ for $y_i \in \mathbb{Z}^n \setminus \{0\}$.

As M is relatively compact, upon passing to a subsequence, we may assume that there exist a sequence $\{\gamma_i\}_{i \in \mathbb{N}} \in \Gamma$ and an element $g \in G$ such that $\lim_i g_i \gamma_i = g$. Since

$$\lim_i g_i \gamma_i (\gamma_i^{-1} y_i) = \lim_i x_i = 0,$$

this implies that

$$\lim_i \gamma_i^{-1} y_i = 0.$$

This is a contradiction as $\gamma_i^{-1} y_i \in \mathbb{Z}^n \setminus \{0\}$.

To show that (ii) implies (i), let $\Sigma_{t,u}$ be a Siegel domain in G. By Theorem 1.7, we can choose t, u such that $\mathcal{L} = \Sigma_{t,u}(L_0)$. So, $M = M'(L_0)$ for a subset M' of $\Sigma_{t,u}$. We claim that there are constants $\alpha, \beta > 0$ such that

$$\alpha \le (a_g)_i \le \beta, \qquad \forall g \in M', \quad i = 1, \ldots, n.$$

(For $g \in G$, recall that a_g denotes the A-component of g in the Iwasawa decomposition and $(a_g)_i$ the i-th diagonal element of a_g.) The claim implies that M' and, hence, M is relatively compact.

Now, by hypothesis, there exists $c > 0$ such that

$$\|g(x)\| \ge c, \qquad \forall x \in \mathbb{Z}^n \setminus \{0\}, \quad g \in M'.$$

Hence,

$$(a_g)_1 = \|g(e_1)\| \ge c, \qquad \forall g \in M'.$$

As $a_g \in A_t$, this implies that there exists $\alpha > 0$ such that

$$(a_g)_i \ge \alpha, \qquad \forall g \in M', \quad i = 1, \ldots, n.$$

Since $(a_g)_1 (a_g)_2 \cdots (a_g)_n = \det a_g = 1$, it follows that, for some constant $\beta > 0$,

$$(a_g)_i \le \beta, \qquad \forall g \in M', \quad i = 1, \ldots, n.$$

This finishes the proof. $\qquad\qquad\qquad\qquad\qquad\qquad\qquad\qquad\qquad\square$

§4 Reduction of Positive Definite Quadratic Forms

A third application of Siegel sets is a classical result about reduction of positive definite quadratic forms.

Let $F : \mathbb{R}^n \times \mathbb{R}^n \to \mathbb{R}$ be a symmetric bilinear form. Then F may be identified with the symmetric $n \times n$ matrix

$$\sigma = (F(e_i, e_j))_{1 \leq i,j \leq n}.$$

The *quadratic form* associated with F is the function $Q : \mathbb{R}^n \to \mathbb{R}$ defined by

$$Q(x) = F(x,x), \qquad \forall x \in \mathbb{R}^n.$$

As

$$F(x,y) = \frac{1}{2}(Q(x+y) - Q(x) - Q(y)),$$

F is completely determined by Q.

Call Q (or F) *positive definite* if

$$Q(x) > 0, \qquad \forall x \in \mathbb{R}^n \setminus \{0\}.$$

This is the case if and only if the corresponding symmetric matrix σ is positive and nonsingular.

Let $\mathcal{H}$ be the set of all positive definite quadratic forms on $\mathbb{R}^n$, identified with the set of all symmetric positive non-singular matrices. The group $G = \mathrm{GL}(n, \mathbb{R})$ acts from the right on $\mathcal{H}$ by the rule

$$\sigma \to g^t \sigma g, \qquad \forall g \in G, \ \sigma \in \mathcal{H}.$$

If $x \mapsto Q(x)$ is the quadratic form corresponding to σ, then $x \mapsto Q(gx)$ is the quadratic form corresponding to $g^t \sigma g$.

Let $\mathcal{H}^{(1)}$ be the space of the forms with determinant 1. The following is easy to prove and left as an exercise.

4.1 Lemma.
(i) $\mathrm{GL}(n, \mathbb{R})$ *acts transitively on* $\mathcal{H}$ *and the stabilizer of the identity* I *is the orthogonal group* $\mathrm{O}(n, \mathbb{R})$.
(ii) $\mathrm{SL}(n, \mathbb{R})$ *acts transitively on* $\mathcal{H}^{(1)}$ *and the stabilizer of the identity* I *is* $\mathrm{SO}(n, \mathbb{R})$.

Therefore, $\mathcal{H}$ may be identified with $\mathrm{O}(n, \mathbb{R}) \backslash \mathrm{GL}(n, \mathbb{R})$ and $\mathcal{H}^{(1)}$ with $\mathrm{SO}(n, \mathbb{R}) \backslash \mathrm{SL}(n, \mathbb{R})$.

Two forms from $\mathcal{H}$ (or $\mathcal{H}^{(1)}$) are usually called *integrally equivalent* if they belong to the same $\mathrm{GL}(n, \mathbb{Z})$-orbit (or $\mathrm{SL}(n, \mathbb{Z})$-orbit). A classical problem is to find *reduced forms*, that is, to choose a reasonable set of representatives for these equivalence classes. This is an important problem as, for instance, two integrally equivalent forms take the same set of values on the lattice $\mathbb{Z}^n$. The following is a classical result due to H. Minkowski.

4.2 Theorem (Minkowski). *Let*

$$\Sigma'_{t,u} = \{n^t an|\ a \in A_t, n \in N_u\}.$$

Then, for $t \geq 4/3$ *and* $u \geq 1/2$,

$$\mathcal{H}^{(1)} = \Sigma'_{t,u}\mathrm{SL}(n,\mathbb{Z}),$$

and $\mathcal{H}^{(1)}/\mathrm{SL}(n,\mathbb{Z})$ *has finite volume (with respect to an invariant measure on* $\mathrm{SO}(n,\mathbb{R})\backslash\mathrm{SL}(n,\mathbb{R})$ *).*

Proof Let $g = kan \in \Sigma_{t,u}$. Then

$$g^t g = n^t a^t k^t kan = n^t a^2 n \in \Sigma'_{t^2,u}.$$

For $t \geq 2/\sqrt{3}$ and $u \geq 1/2$, one has $\mathrm{SL}(n,\mathbb{R}) = \Sigma_{t,u}\mathrm{SL}(n,\mathbb{Z})$, by Theorem 1.7. This proves the claim. $\quad\square$

4.3 Remark. An analogous result is valid for forms in $\mathcal{H}$. Indeed, Siegel sets in $\mathrm{GL}(n,\mathbb{R})$ are defined similarly by

$$\Sigma_{t,u} = KA_t N_u,$$

where $K = \mathrm{O}(n,\mathbb{R})$ and A is the group of the diagonal matrices with positive diagonal entries. Then

$$\mathrm{GL}(n,\mathbb{R}) = \Sigma_{t,u}\mathrm{GL}(n,\mathbb{Z}),$$

and

$$\mathcal{H} = \Sigma'_{t,u}\mathrm{GL}(n,\mathbb{Z}).$$

However, notice that $\mathrm{GL}(n,\mathbb{R})/\mathrm{GL}(n,\mathbb{Z})$ (and $\mathcal{H}/\mathrm{GL}(n,\mathbb{Z})$) has infinite volume.

As another application of Siegel sets, here is a classical result due to C. Hermite.

4.4 Theorem (Hermite). *Let* Q *be a positive definite quadratic form on* $\mathbb{R}^N$. *Then there exists* $x \in \mathbb{Z}^n \setminus \{0\}$ *such that*

$$Q(x) \leq (4/3)^{\frac{n-1}{2}}(\det Q)^{1/n}.$$

Proof Let $\sigma = gg^t$, $g \in \mathrm{GL}(n,\mathbb{R})$, be the positive matrix corresponding to Q. So,

$$Q(x) = x^t \sigma x = \|gx\|^2, \qquad \forall x \in \mathbb{R}^n.$$

Replacing Q by $(\det Q)^{-\frac{1}{n}}Q$, we may assume that $g \in \mathrm{SL}(n,\mathbb{R})$. By Theorem 1.7, there exist $g' \in \Sigma_{2/\sqrt{3},1/2}$ and $\gamma \in \mathrm{SL}(n,\mathbb{Z})$ such that $g\gamma = g'$.

Set $x = \gamma e_1 \in \mathbb{Z}^n \setminus \{0\}$. Let $g' = kan$ be the Iwasawa decomposition of g'. Then, denoting by $a_1, \ldots, a_n$ the diagonal entries of a,

$$\|gx\| = \|g\gamma(e_1)\| = \|g'e_1\| = \|ae_1\| = a_1.$$

As $g' \in \Sigma_{2/\sqrt{3}, 1/2}$,

$$a_1^n \le (2/\sqrt{3})^{n-1} a_1 a_2^{n-1} \le \cdots \le (2/\sqrt{3})^{n(n-1)/2} a_1 a_2 \cdots a_n = (2/\sqrt{3})^{n(n-1)/2},$$

and, hence,

$$Q(x) = \|gx\|^2 = a_1^2 \le (4/3)^{\frac{n-1}{2}}.$$

$\square$

§5 Margulis' Lemma

A one-parameter subgroup $\{u(t)\}_{t \in \mathbb{R}}$ of $GL(n, \mathbb{R})$ is called *unipotent* if there exists a nilpotent endomorphism τ on $\mathbb{R}^n$ with

$$u(t) = \exp(t\tau) \qquad \forall t \in \mathbb{R}.$$

Clearly, any such subgroup is contained in $SL(n, \mathbb{R})$.

In its original version, Margulis' Lemma is the following statement ([Ma5]).

5.1 Theorem (Margulis' Lemma). *Let* $n \ge 2$. *Let* $\{u_t\}_{t \in \mathbb{R}}$ *be a unipotent one-parameter group of* $SL(n, \mathbb{R})$ *and let* $x \in SL(n, \mathbb{R})/SL(n, \mathbb{Z})$. *Then* $u_t x$ *does not tend to infinity as* $t \to +\infty$. *That is, there exists a compact subset* $K \subseteq SL(n, \mathbb{R})/SL(n, \mathbb{Z})$ *such that the set*

$$\{t \ge 0 \,|\, u_t x \in K\}$$

is unbounded.

Margulis' Lemma may be rephrased in terms of lattices in $\mathbb{R}^n$. Recall (see the beginning of Section 3) that the set $\mathcal{L}$ of all unimodular lattices in $\mathbb{R}^n$ may be identified with the homogeneous space $SL(n, \mathbb{R})/SL(n, \mathbb{Z})$, via the natural action of $SL(n, \mathbb{R})$ on $\mathcal{L}$. By Mahler's criterion (Theorem 3.2), for any $\delta > 0$, the set

$$\{gSL(n, \mathbb{Z}) \,|\, \|gp\| \ge \delta, \forall p \in \mathbb{Z}^n, p \ne 0\}$$

is a compact subset of $SL(n, \mathbb{R})/SL(n, \mathbb{Z})$. So, Margulis' Lemma has the following equivalent formulation.

5.2 Theorem (Margulis' Lemma). *Let* $n \ge 2$. *Let* $\{u_t\}_{t \in \mathbb{R}}$ *be a unipotent one-parameter group of* $SL(n, \mathbb{R})$. *For any lattice* Λ *in* $\mathbb{R}^n$, *there exists* $\delta > 0$ *such that the set*

$$\{t \ge 0 \,|\, \|u_t p\| \ge \delta, \forall p \in \Lambda, p \ne 0\}$$

is unbounded.

For later applications in Chapter VI, we shall need a "uniform" version of Margulis' Lemma (see Theorem 5.12 and Theorem 5.17 below).

We shall give complete proofs, following the elementary exposition in [DM3].

The arguments depend on the behaviour of a certain function defined on all discrete subgroups of $\mathbb{R}^n$ and which will now be introduced.

Let $\mathbb{R}^n$ be equipped with the inner product which makes the standard basis $e_1, \ldots, e_n$ an orthonormal basis. Let Δ be a discrete subgroup of $\mathbb{R}^n$, and let $\Delta_{\mathbb{R}}$ be the subspace spanned by Δ. There exists a basis $\{x_1, \ldots, x_r\}$ of $\Delta_{\mathbb{R}}$ which generates the group Δ, and Δ is a lattice in $\Delta_{\mathbb{R}}$. The number

$$d(\Delta) = \mathrm{vol}(\Delta_{\mathbb{R}}/\Delta)$$

is called the *determinant* of Δ. (The measure on $\Delta_{\mathbb{R}}$ is determined by the inner product inherited from $\mathbb{R}^n$.) Observe that $d(\Delta) = |\det \tau|$ for any linear transformation τ on $\Delta_{\mathbb{R}}$ such that $\tau^{-1}x_1, \ldots, \tau^{-1}x_r$ is an orthonormal base of $\Delta_{\mathbb{R}}$.

Some properties of the function d will be crucial for what follows. We shall frequently use the following explicit formulae for $d^2(\Delta)$.

5.3 Lemma. *Let Δ be a discrete subgroup of $\mathbb{R}^n$, generated by linearly independent vectors $x_1, \ldots, x_r$. Then*

(i) $d^2(\Delta)$ *is equal to the determinant of the $r \times r$ matrix $((\langle x_i, x_j \rangle))_{1 \le i,j \le r}$.*

(ii) $d^2(\Delta)$ *is equal to the sum of squares of the determinants of all $r \times r$ minors of A, where A is the $n \times r$ matrix with columns $x_1, \ldots, x_r$.*

Proof (i) Let τ be a linear transformation of $\Delta_{\mathbb{R}}$ such that

$$\tau^{-1}x_1, \ldots, \tau^{-1}x_r$$

is an orthonormal base of $\Delta_{\mathbb{R}}$. Then

$$d(\Delta)^2 = (\det \tau)^2 = \det(\tau^t \tau) = \det((\langle x_i, x_j \rangle)_{1 \le i,j \le r}).$$

By (i), one has $d^2(\Delta) = \det A^t A$. So, (ii) is a consequence of the following lemma. $\square$

We need some notation. For an $n \times r$ matrix A with $r \le n$ and rows $a_1, \ldots, a_n$, we denote by $A_{i_1, \ldots, i_r}$ the $r \times r$ matrix with rows $a_{i_1}, \ldots, a_{i_r}$.

5.4 Lemma. *Let A, B be $n \times r$ matrices, $r \le n$. Then*

$$\det(A^t B) = \sum_{1 \le i_1 < \cdots < i_r \le n} \det A_{i_1, \ldots, i_r} \det B_{i_1, \ldots, i_r}.$$

Proof Consider the two mappings

$$\Phi : X \mapsto \det(X^t A)$$

$$\Psi : X \mapsto \sum_{1 \le i_1 < \ldots < i_r \le n} \det A_{i_1,\ldots,i_r} \det X_{i_1,\ldots,i_r}$$

as functions of the columns of the $n \times r$ matrix X. Viewed this way, Φ and Ψ are multilinear on $(\mathbb{R}^n)^r$.

It is clear that Φ and Ψ agree on matrices X of the form

$$X = (e_{j_1}, \ldots, e_{j_r})$$

for all $1 \le j_1, \ldots, j_r \le n$. Hence, they agree everywhere. $\square$

5.5 Lemma. *Let Δ be a discrete subgroup of $\mathbb{R}^n$, and let $x \in \mathbb{R}^n, x \notin \Delta_{\mathbb{R}}$. Then*

$$d(\Delta') \le \|x\| d(\Delta),$$

where Δ' is the (discrete) subgroup generated by Δ and x.

Proof Let $x_1, \ldots, x_r$ be linearly independent generators of Δ. Then, by Lemma 5.3 (ii), $d(\Delta')^2$ is the sum of squares of the determinants of the $(r+1) \times (r+1)$ minors of the $n \times (r+1)$ matrix with columns $x_1, \ldots, x_r, x$. Expanding each such determinant according to the last column yields the result. $\square$

5.6 Lemma. *For any $g \in \mathrm{GL}(n, \mathbb{R})$, there exist constants $\alpha, \beta > 0$ such that, for every discrete subgroup Δ of $\mathbb{R}^n$,*

$$\alpha d(\Delta) \le d(g\Delta) \le \beta d(\Delta).$$

Proof Let $x_1, \ldots, x_r$ be linearly independent generators of Δ. Denote by A the $n \times r$ matrix with columns $x_1, \ldots, x_r$, and by B the matrix with columns $gx_1, \ldots, gx_r$.

Recall that, for a matrix C with n rows $c_1, \ldots, c_n$ and for $1 \le i_1 < \cdots < i_r \le n$, we denote by $C_{i_1,\ldots,i_r}$ the matrix consisting of the r rows $c_{i_1}, \ldots, c_{i_r}$. Clearly,

$$B_{i_1,\ldots,i_r} = g_{i_1,\ldots,i_r} A.$$

Hence, by Lemma 5.4,

$$\det B_{i_1,\ldots,i_r} = \sum_{1 \le j_1 < \cdots < j_r \le n} \det g^{i_1,\ldots,i_r}_{j_1,\ldots,j_r} \det A_{j_1,\ldots,j_r},$$

where the $n \times r$ matrix $g^{i_1,\ldots,i_r}$ is the transpose of $g_{i_1,\ldots,i_r}$. Let now α be the square root of

$$\sum_{1 \le i_1 < \cdots < i_r \le n} \sum_{1 \le j_1 < \cdots < j_r \le n} (\det g^{i_1,\ldots,i_r}_{j_1,\ldots,j_r})^2.$$

Then, by the Cauchy–Schwarz inequality,

$$d(g\Delta)^2 = \sum_{1 \le i_1 < \cdots < i_r \le n} (\det B_{i_1,\ldots,i_r})^2 \le \alpha^2 d(\Delta)^2.$$

Applying this inequality to g^{-1} gives the inequality from below .   □

Given a lattice Λ, we shall mainly be interested in the *complete* subgroups of Λ. These are the subgroups of the form $\Lambda \cap W$ for a subspace W of $\mathbb{R}^n$.

5.7 Lemma. *Let Λ be a lattice in $\mathbb{R}^n$. Then, for any $\delta > 0$, the set of all subgroups Δ of Λ with $d(\Delta) < \delta$ is finite.*

Proof   There exists $g \in GL(n, \mathbb{R})$ such that $\Lambda = g\mathbb{Z}^n$. Hence, by the above lemma, we may assume that $\Lambda = \mathbb{Z}^n$. Moreover, as

$$d(\Lambda \cap \Delta_{\mathbb{R}}) \le d(\Delta),$$

for any subgroup Δ of Λ, it suffices to prove the claim for the family of all complete subgroups of Λ.

Let Δ be a complete subgroup of $\mathbb{Z}^n$ with $d(\Delta) < \delta$. Then Δ is generated by r linearly independent vectors $x_1, \ldots, x_r \in \mathbb{Z}^n$. The determinants of the $r \times r$ minors of the matrix with rows $x_1, \ldots, x_r$ are integers. Hence, by Lemma 5.3 (ii) above, there are only finitely many possible values for these determinants. The claim follows now from the next lemma.   □

5.8 Lemma. *Let A, B be $n \times r$ matrices of rank r such that*

$$\det A_{i_1,\ldots,i_r} = \det B_{i_1,\ldots,i_r}$$

for all $1 \le i_1 < \cdots < i_r \le n$. Then the columns of A and B span the same linear subspace of $\mathbb{R}^n$.

Proof   Let $a_1, \ldots, a_r$ and $b_1, \ldots, b_r$ be the columns of A and B. For $1 \le i \le r$, let $B^{(i)}$ be the $n \times (r+1)$ matrix with columns $a_1, \ldots, a_r, b_i$. Let $C^{(i)}$ be the $n \times (r+1)$ matrix with columns $b_1, \ldots, b_r, b_i$.

Expanding the determinant of $B^{(i)}_{i_1,\ldots,i_{r+1}}$ according to the last column, the assumption on A and B shows that

$$\det B^{(i)}_{i_1,\ldots,i_{r+1}} = \det C^{(i)}_{i_1,\ldots,i_{r+1}} = 0,$$

for all $1 \le i_1 < \cdots < i_{r+1} \le n + 1$. This shows that $a_1, \ldots, a_r, b_i$ are linearly dependent for all $1 \le i \le r$.   □

The following is a simple but crucial fact.

5.9 Lemma. *Let Δ be a discrete subgroup of $\mathbb{R}^n$, and let $\{u_t\}_{t\in\mathbb{R}}$ be a unipotent one-parameter group of $\mathrm{SL}(n,\mathbb{R})$. Then $d^2(u_t\Delta)$ is a polynomial in t, of degree at most $2n(n-1)$.*

Proof Write $u_t = \exp(tA)$ for a nilpotent matrix A. Then $A^n = 0$. Hence, for any $x \in \mathbb{R}^n$, the coordinates of $u_t x$ are polynomials in t of degree at most $n-1$. Let $x_1, \ldots, x_r$ be linearly independent generators of Δ. Then $\langle u_t x_i, u_t x_j \rangle$ is a polynomial in t of degree at most $2(n-1)$. As $d^2(u_t\Delta) = \det(\langle u_t x_i, u_t x_j \rangle)_{1\le i,j\le n}$, this proves the claim. $\square$

Some growth properties of the polynomial $d^2(u_t\Delta)$ will play an important role. We state such properties in the following elementary lemmas, which are in the spirit of Chap. IV, Lemma 4.3.

Let $\mathcal{P}_m$ denote the space of all polynomials on $\mathbb{R}$ of degree at most m, and let $\mathcal{P}_m^+$ be the subset of all $P \in \mathcal{P}_m$ with $P(x) \ge 0$ for all $x \in \mathbb{R}$. For any (non-empty) interval $[a,b]$,

$$\|P\|_{[a,b]} := \max\{|P(t)| \,|\, t \in [a,b]\}, \qquad P \in \mathcal{P}_m$$

is a norm on $\mathcal{P}_m$. The arguments to follow rely on the compactness of closed balls in $\mathcal{P}_m$.

5.10 Lemma. *Fix $m \in \mathbb{N}$ and $\lambda \in \mathbb{R}$ with $\lambda > 1$. There exists $\varepsilon = \varepsilon(m,\lambda) > 0$ with the following property. Let $a < b$, $\alpha > 0$, and let $P \in \mathcal{P}_m^+$ be such that $\|P\|_{[a,b]} \ge \alpha$ and $P(b) < \alpha\varepsilon$. Then there exists $t \in [b, a+\lambda(b-a)]$ such that $P(t) = \alpha\varepsilon$.*

Proof As $\mathcal{P}_m$ is a finite-dimensional vector space, the norms $\|P\|_{[0,1]}$ and $\|P\|_{[1,\lambda]}$ are equivalent. In particular, there exists $\varepsilon > 0$ with

$$\varepsilon\|P\|_{[0,1]} \le \|P\|_{[1,\lambda]}, \qquad \forall P \in \mathcal{P}_m.$$

Let $P \in \mathcal{P}_m^+$ be such that $\|P\|_{[a,b]} \ge \alpha$ and $P(b) < \alpha\varepsilon$. Then, for the polynomial $\widetilde{P}$ in $\mathcal{P}_m$ defined by $\widetilde{P}(t) = P(a+t(b-a))$, one has

$$\|\widetilde{P}\|_{[0,1]} = \|P\|_{[a,b]} \quad \text{and} \quad \|\widetilde{P}\|_{[1,\lambda]} = \|P\|_{[b,a+\lambda(b-a)]}.$$

The above inequality applied to $\widetilde{P}$ shows that

$$\|P\|_{[b,a+\lambda(b-a)]} \ge \varepsilon\alpha.$$

As $P(b) < \varepsilon\alpha$, the claim follows. $\square$

5.11 Lemma. *Fix $m \in \mathbb{N}$ and $\mu \in \mathbb{R}$ with $\mu > 1$. There exist real numbers $\varepsilon_1 = \varepsilon_1(m,\mu) > 0$ and $\varepsilon_2 = \varepsilon_2(m,\mu) > 0$ with the following property. Let $a < b$, $\alpha > 0$, and let $P \in \mathcal{P}_m^+$ be such that $\|P\|_{[a,b]} \le \alpha$*

and $P(b) = \alpha$. Then there exists $0 \le i \le m$ such that

$$\alpha\varepsilon_1 \le P(t) \le \alpha\varepsilon_2 \qquad \forall t \in [a + \mu^{2i+1}(b-a), a + \mu^{2i+2}(b-a)].$$

Proof Let $I_i = [\mu^{2i+1}, \mu^{2i+2}]$, $0 \le i \le m$. The I_i's are $m+1$ pairwise disjoint intervals. The set X of all polynomials $P \in \mathcal{P}_m$ with $\|P\|_{[0,1]} \le 1$ and $P(1) = 1$ is compact.

The mapping

$$\Phi : \prod_{i=0}^{m} I_i \times X \to \mathbb{R}, \qquad ((t_0, \dots, t_m), P) \mapsto \sum_{i=0}^{m} |P(t_i)|$$

is continuous and has strictly positive values, since a non-zero polynomial in $\mathcal{P}_m$ has at most m roots. Hence, there exists $\delta_1, \delta_2 > 0$ such that

$$\delta_1 \le \Phi(\xi) \le \delta_2$$

for all ξ in the compact set $\prod_{i=0}^{m} I_i \times X$. Set

$$\varepsilon_1 = \delta_1/m + 1, \qquad \varepsilon_2 = \delta_2.$$

Then, for each $P \in X$, there exists $0 \le i \le m$ such that

$$\varepsilon_1 \le |P(t)| \le \varepsilon_2 \qquad \forall t \in I_i.$$

Let $P \in \mathcal{P}_m^+$ be such that $\|P\|_{[a,b]} \le \alpha$ and $P(b) = \alpha$. Define

$$\widetilde{P}(t) = \frac{1}{\alpha}P(a + t(b-a)).$$

Then $\widetilde{P} \in X$. Apply now the above inequality to $\widetilde{P}$. $\square$

The following is the main result of this section.

5.12 Theorem (Margulis' Lemma – uniform version). *Let $n \ge 2$ and let $\sigma > 0$. Then there exists $\delta > 0$ such that, for any lattice Λ in $\mathbb{R}^n$, for any unipotent one-parameter subgroup $\{u_t\}_{t\in\mathbb{R}}$ of $\mathrm{SL}(n,\mathbb{R})$ and any $T \ge 0$, the following holds: either there exists $s \ge T$ such that $\|u_s x\| \ge \delta$ for all $x \in \Lambda, x \ne 0$, or there exists a subgroup Δ of Λ such that $d(u_t\Delta) < \sigma$ for all $t \in [0,T]$.*

For the proof, we shall need several lemmas. The following constants will be fixed throughout the proof.

Notation. Let $\cdot m = 2n^2$ and let $\mu > 1$ be arbitrary. Choose $\lambda > 1$ with

$$\lambda - 1 \le (\mu - 1)/\mu^{2m+2}.$$

Let $0 < \varepsilon = \varepsilon(m, \lambda) < 1$ and $0 < \varepsilon_1 = \varepsilon_1(m, \mu) < 1 < \varepsilon_2 = \varepsilon_2(m, \mu)$ be as in Lemma 5.10 and Lemma 5.11. Set

$$\alpha = \sqrt{\varepsilon}, \qquad \beta_1 = \sqrt{\varepsilon_1}, \qquad \beta_2 = \sqrt{\varepsilon_2}.$$

For a lattice Λ, we denote by the set $\mathcal{S}(\Lambda)$ of all complete subgroups of Λ, partially ordered by the usual inclusion. If M is a totally ordered subset of $\mathcal{S}(\Lambda)$, we define $\mathcal{C}(M, \Lambda)$ to be the set of all subgroups Δ in $\mathcal{S}(\Lambda)$, $\Delta \notin M$, such that $M \cup \{\Delta\}$ is a totally ordered set.

5.13 Lemma. *Let* $\{u_t\}_{t \in \mathbb{R}}$ *be a unipotent one-parameter group of* $\mathrm{SL}(n, \mathbb{R})$, Λ *a lattice in* $\mathbb{R}^n$ *and* S *a totally ordered subset of* $\mathcal{S}(\Lambda)$. *Let* $\tau > 0$ *and* $T \geq 0$ *be such that, for each* $\Phi \in \mathcal{C}(S, \Delta)$, *there exists* $t \in [0, T]$ *such that* $d(u_t \Phi) \geq \tau$. *Then either* $d(u_T \Phi) \geq \alpha \tau$ *for all* $\Phi \in \mathcal{C}(S, \Lambda)$ *or there exist* $\Delta \in \mathcal{C}(S, \Lambda)$ *and* $T_1 \in [T, (2 - \mu^{-1})T]$ *such that the following holds:*
(i) $\tau \alpha \beta_1 \leq d(u_t \Delta) \leq \tau \alpha \beta_2$ *for all* $t \in [T_1, T + \mu(T_1 - T)]$;
(ii) *For each* $\Phi \in \mathcal{C}(S, \Lambda)$, *there exists* $t \in [T, T_1]$ *such that* $d(u_t \Phi) \geq \alpha \tau$.

Proof We may assume that the set $\mathcal{F}$ of all $\Phi \in \mathcal{C}(S, \Delta)$ such that

$$d(u_T \Phi) < \alpha \tau$$

is non-empty. By Lemma 5.7, $\mathcal{F}$ has only finitely many elements $\Phi_1, \ldots, \Phi_r$.

The positive polynomial $d(u_t \Phi_j)^2$ of degree at most $2n(n-1)$ satisfies the hypotheses of Lemma 5.10, for each $1 \leq j \leq r$. Hence, there exists $t_j \in [T, \lambda T]$ with $d(u_{t_j} \Phi_j) = \alpha \tau$ and $d(u_t \Phi_j) \leq \alpha \tau$ for all $T \leq t \leq t_j$.

Let $t_k = \max\{t_j | 1 \leq j \leq r\}$ and $\Delta = \Phi_k$. By Lemma 5.11, there exists $0 \leq i \leq m$ such that

$$\tau \alpha \beta_1 \leq d(u_t \Delta) \leq \tau \alpha \beta_2, \qquad \forall t \in [T_1, T_2],$$

where $T_1 = T + \mu^{2i+1}(t_k - T)$ and $T_2 = T + \mu^{2i+2}(t_k - T)$. As $t_k \in [T, \lambda T]$ and $\lambda - 1 \leq (\mu - 1)/\mu^{2m+2}$,

$$T + \mu(T_1 - T) = T_2 \leq T + \mu^{2m+2}(t_k - T) \leq T + \mu^{2m+2}(\lambda - 1) \leq \mu T,$$

and hence $T_1 \in [T, (2 - \mu^{-1})T]$. This shows that condition (i) in the lemma is satisfied. As $T \leq t_j \leq t_k \leq T_1$ for all $1 \leq j \leq r$, condition (ii) is also satisfied. $\qquad\square$

The following lemma is the crucial step in the proof of Theorem 5.12.

5.14 Lemma. *Let* $n \geq 2$, *let* $\{u_t\}_{t \in \mathbb{R}}$ *be a unipotent one-parameter group of* $\mathrm{SL}(n, \mathbb{R})$ *and let* Λ *be a lattice in* $\mathbb{R}^n$. *Let* $\tau > 0$ *and* $T \geq 0$ *be such that, for each* $\Phi \in \mathcal{S}(\Lambda)$, *there exists* $t \in [0, T]$ *such that* $d(u_t \Phi) \geq \tau$. *Then there exists a totally ordered subset* M *of* $\mathcal{S}(\Lambda)$, *and* $R \in [T, \mu T]$ *with the following properties:*
(i) $\tau \alpha^n \beta_1 \leq d(u_R \Phi) \leq \tau \alpha \beta_2$ *for all* $\Phi \in M$;
(ii) $d(u_R \Phi) \geq \alpha^n \tau$, *for all* $\Phi \in \mathcal{C}(M, \Lambda)$.

Proof We assume that $n = 3$. This case already contains all the features of the general case. It will be clear how the inductive procedure continues for arbitrary n.

If $d(u_T\Phi) \geq \alpha\tau$, for all $\Phi \in \mathcal{S}(\Lambda)$, then set

$$M = \emptyset \quad \text{and} \quad R = T.$$

If not, Lemma 5.13 above, applied to $S = \emptyset$, shows that there exist $\Delta_1 \in \mathcal{S}(\Lambda)$ and $T_1 \in [T, (2 - \mu^{-1})T]$ such that

$$\tau\alpha\beta_1 \leq d(u_t\Delta_1) \leq \tau\alpha\beta_2$$

for all $t \in [T_1, T + \mu(T_1 - T)]$ and, for each $\Phi \in \mathcal{S}(\Lambda)$, there exists $t \in [T, T_1]$ such that $d(u_t\Phi) \geq \alpha\tau$.

If $d(u_{T_1}\Phi) \geq \alpha^2\tau$, for all $\Phi \in \mathcal{C}(\Delta_1, \Lambda)$, then set

$$M = \{\Delta_1\} \quad \text{and} \quad R = T_1.$$

Otherwise, apply Lemma 5.13 to $S = \{u_T(\Delta_1)\}$ and to $T_1 - T$ and $\alpha\tau$ instead of T and τ. This shows that there exist $\Delta_2 \in \mathcal{C}(\Delta_1, \Lambda)$ and

$$T_2 \in [T_1, (2 - \mu^{-1})(T_1 - T) + T]$$

such that

$$\tau\alpha^2\beta_1 \leq d(u_t\Delta_2) \leq \tau\alpha^2\beta_2$$

for all $t \in [T_2, T_1 + \mu(T_2 - T_1)]$ and, for each $\Phi \in \mathcal{C}(\Delta_1, \Lambda)$, there exists $t \in [T_1, T_2]$ such that $d(u_t\Phi) \geq \alpha^2\tau$. Observe that, since

$$T_2 \leq (2 - \mu^{-1})(T_1 - T) + T,$$

one has

$$T_1 + \mu(T_2 - T_1) \leq T_1 + \mu[(2 - \mu^{-1})(T_1 - T) + T - T_1] = T + \mu(T_1 - T).$$

Now, if $d(u_{T_2}\Phi) \geq \alpha^3\tau$, for all $\Phi \in \mathcal{C}(\{\Delta_1, \Delta_2\}, \Lambda)$, then set

$$M = \{\Delta_1, \Delta_2\} \quad \text{and} \quad R = T_2.$$

Otherwise, apply again Lemma 5.13 to $S = \{u_{T_1}\Delta_1, u_{T_1}\Delta_2\}$ and to $T_2 - T_1$ and $\alpha^2\tau$ instead of T and τ. This shows that there exist $\Delta_3 \in \mathcal{C}(S, \Lambda)$ and

$$T_3 \in [T_2, (2 - \mu^{-1})(T_2 - T_1) + T_1]$$

such that

$$\tau\alpha^3\beta_1 \leq d(u_t\Delta_3) \leq \tau\alpha^3\beta_2$$

for all $t \in [T_3, T_2 + \mu(T_3 - T_1)]$. Since

$$T_3 \leq (2 - \mu^{-1})(T_2 - T_1) + T_1,$$

one has, as above,

$$T_2 + \mu(T_3 - T_2) \leq T_1 + \mu(T_2 - T_1).$$

Take now

$$M = \{\Delta_1, \Delta_2, \Delta_3\} \quad \text{and} \quad R = T_3.$$

Thus (i) is satisfied. Observe that, since $n = 3$, the set $C(M, \Lambda)$ is empty, and condition (ii) is trivially satisfied. The procedure comes to an end. It is clear how one has to continue for arbitrary n.   □

At last, we are now ready to prove Theorem 5.12.

Proof of Theorem 5.12   Fix an arbitrary real number $\mu > 1$, and let α, β_1, β_2 be as before Lemma 5.11. Recall that $0 < \alpha < 1$ and $0 < \beta_1 < 1 < \beta_2$.

Let T, σ be as in the statement of the theorem. Set $\tau = \min\{\sigma, 1\}$, and let

$$\delta = \alpha^n \beta_1 \beta_2^{-1} \tau.$$

Suppose that Λ has no non-zero subgroup Δ with $d(u_t \Delta) < \tau$ for all $t \in [0, T]$. Let M be a totally ordered subset of $\mathcal{S}(\Lambda)$ and $R \in [T, \mu T]$ be as in the previous Lemma 5.14. We show that, for all $x \in \Lambda$, $x \neq 0$,

$$\|u_R x\| \geq \delta.$$

Clearly, it is enough to prove this inequality for a primitive element $x \in \Lambda$. (Recall that x is primitive if $x = ky$ for $y \in \Lambda$, $k \in \mathbb{Z}$ implies that $k = 1$ or $k = -1$.) Let $\Delta = \langle x \rangle$ be the subgroup generated by x. Then $\Delta \in \mathcal{S}(\Lambda)$. Two cases may occur:

• *First case*: $x \in \Phi$ for all $\Phi \in M$. Then $\Delta \in M$ or $\Delta \in C(M, \Lambda)$, and hence, by the previous lemma,

$$\|u_R x\| \geq d(u_R \Delta) \geq \alpha^n \beta_1 \tau \geq \delta.$$

• *Second case*: $x \notin \Phi$ for some $\Phi \in M$. Let Ψ be the largest subgroup in M with $x \notin \Psi$. Let $\Delta' \in \mathcal{S}(\Lambda)$ be the *complete* subgroup of Λ generated by x and Ψ.

Observe that, if $\Phi \in M$ contains properly Ψ, then Φ contains x. Hence, either $\Delta' \in C(M, \Lambda)$ or $\Delta' \in M$. Thus, by Lemma 5.14,

$$d(u_R \Delta') \geq \alpha^n \beta_1 \tau \qquad \text{and} \qquad d(u_R \Psi) \leq \beta_2.$$

Since, by Lemma 5.5,

$$d(u_R \Delta') \leq \|u_R x\| d(u_R \Psi),$$

this implies that

$$\|u_R x\| \geq \alpha^n \beta_1 \beta_2^{-1} \tau = \delta,$$

and the proof is complete.   □

5.15 Corollary.   *For every $\sigma > 0$, there exists $\delta > 0$ such that, for any unipotent one-parameter group $\{u_t\}$ of $\mathrm{SL}(n, \mathbb{R})$ and any lattice Λ in $\mathbb{R}^n$, the following holds: either the set*

$$\{t \geq 0 \mid \|u_t x\| \geq \delta, \forall x \in \Lambda, x \neq 0\}$$

is unbounded or there exists a non-zero subgroup Δ *of* Λ *such that* $\Delta_{\mathbf{R}}$ *is invariant under* $\{u_t\}$ *and* $d(u_t\Delta) = d(\Delta) < \sigma$ *for all* $t \in \mathbb{R}$.

Proof Let $\delta > 0$ be as in Theorem 5.12, and suppose that the first possibility does not hold. Then, using Lemma 5.7, it is clear that there exists a non-zero subgroup Δ of Λ such that $d(u_t\Delta) < \sigma$ for all $t \in \mathbb{R}$. Since $t \mapsto d(u_t\Delta)^2$ is a positive polynomial, it has to be constant, that is,

$$d(u_t\Delta) = d(\Delta), \qquad \forall t \in \mathbb{R}.$$

Apply now the following lemma. $\qquad\qquad\qquad\qquad\qquad\qquad\qquad$ □

5.16 Lemma. *Let* Δ *be a discrete subgroup of* $\mathbb{R}^n$, *and let* $\{u_t\}_{t\in\mathbb{R}}$ *be a unipotent one-parameter group of* $\mathrm{SL}(n,\mathbb{R})$ *such that* $d(u_t\Delta) = d(\Delta)$ *for all* $t \in \mathbb{R}$. *Then* $\Delta_{\mathbf{R}}$ *is invariant under* u_t *for all* $t \in \mathbb{R}$.

Proof Let $x_1, \ldots, x_r$ be linearly independent generators of Δ. The positive polynomial $d^2(u_t\Delta)$ is equal to the sum of squares of the determinants of all $r \times r$ minors of the matrix with columns $u_t x_1, \ldots, u_t x_r$ (see Lemma 5.3). Any such determinant is a polynomial in t. As the sum of their squares is constant, all these determinants must be constant. Lemma 5.8 shows that $u_t x_1, \ldots, u_t x_r$ span the same linear subspace as $x_1, \ldots, x_r$, for all $t \in \mathbb{R}$. $\qquad\qquad\qquad\qquad\qquad$ □

We reformulate the above results in terms of orbits of unipotent groups acting on $\mathrm{SL}(n,\mathbb{R})/\mathrm{SL}(n,\mathbb{Z})$.

It is clear that Theorem 5.12 and Corollary 5.15 may be rephrased as follows.

5.17 Theorem (Margulis' Lemma – uniform version). *Let* $n \geq 2$ *and let* $\sigma > 0$. *Then there exists a compact subset* K *of* $\mathrm{SL}(n,\mathbb{R})/\mathrm{SL}(n,\mathbb{Z})$ *such that, for any* $x = g\mathrm{SL}(n,\mathbb{Z}) \in \mathrm{SL}(n,\mathbb{R})/\mathrm{SL}(n,\mathbb{Z})$ *and for any unipotent one-parameter subgroup* $\{u_t\}_{t\in\mathbb{R}}$ *of* $\mathrm{SL}(n,\mathbb{R})$, *the following holds: either the set* $\{t \geq 0 \,|\, u_t x \in K\}$ *is unbounded or there exists a non-zero subgroup* Δ *of* $\mathbb{Z}^n$ *such that the subspace* $\Delta_{\mathbf{R}}$ *is invariant under the one-parameter group* $\{g^{-1}u_t g\}_{t\in\mathbb{R}}$ *and such that* $d(u_t g\Delta) = d(g\Delta) < \sigma$ *for all* $t \in \mathbb{R}$.

It is now easy to deduce Theorem 5.1, the original version of Margulis' Lemma, from the previous theorem.

Proof of Theorem 5.1 Write $x = g\mathrm{SL}(n,\mathbb{Z})$ for some $g \in \mathrm{SL}(n,\mathbb{R})$, and let $\Lambda = g\mathbb{Z}^n$. By Lemma 5.7, there exists $\sigma > 0$ such that $d(\Delta) > \sigma$ for all subgroups Δ of Λ. Now apply Theorem 5.12. $\qquad\qquad$ □

Notes

Our treatement of Siegel sets and Mahler's criterion is based on the monographs [Bor] and [Rag].

The first proof for finiteness of the volume of $SL(n, \mathbb{R})/SL(n, \mathbb{Z})$ is due to H. Minkowski [Mi]. The precise computation of this volume may be found in [Sg1], Lecture XV.

Mahler's original proof of his compactness criterion is in [Mah1]. Other proofs of Mahler's criterion have been given by Chabauty and Macbeath and Swierczkowski (see [Bo2], Chap. 8, §5).

Margulis' Lemma was proved by G. A. Margulis in [Ma5] and used in one of his proofs of the arithmeticity theorem (see [Ma6]). According to [Ma5], it has been conjectured by Pyatetskii-Shapiro. The proof we give is based on the elementary exposition in [DM3].

A quantitative version of Margulis' Lemma, valid for arbirary lattices of $SL(2, \mathbb{R})$, was given in Chap. IV, Theorem 4.2. Actually, this quantitative version is true for any Lie group G, any lattice Γ in G and any Ad-unipotent one-parameter subgroup of G (for the notion of Ad-unipotence, see Chap. VI, §6). This result is due essentially to S. G. Dani (see [Da3], Theorem 3.5 and [DM4], Theorem 6.1) . Let us mention the following application, also due to Dani, of this result. Let Γ be a lattice in a Lie group G, and let U be an Ad-unipotent one-parameter subgroup of G. Then every ergodic locally finite U-invariant measure on $\Gamma \setminus G$ is finite (see [Da2], [Da7]).

Chapter VI

An Application to Number Theory:

Oppenheim's Conjecture

In this chapter, we shall be concerned with Oppenheim's conjecture on values of indefinite irrational quadratic forms at integer points.

Let Q be a real non-degenerate quadratic form $\mathbb{R}^n$. We are interested in the set of values $Q(\mathbb{Z}^n)$ of Q on $\mathbb{Z}^n$, and especially in the question whether $Q(\mathbb{Z}^n)$ is dense in $\mathbb{R}$. It is obvious that if Q is definite (positive definite or negative definite), then $Q(\mathbb{Z}^n)$ is a discrete subset of $\mathbb{R}$. (Indeed, in this case, there exists a constant $C > 0$ such that $|Q(x)| \geq C\|x\|$ for all $x \in \mathbb{R}^n$.) It is also clear that $Q(\mathbb{Z}^n)$ is a discrete set if Q is a *rational form*, that is, if Q is a multiple of a form with rational coefficients. In its strong form (see below) Oppenheim's conjecture says that, if Q is an indefinite *irrational* (that is, Q has two coefficients with an irrational ratio) in $n \geq 3$ variables, then $Q(\mathbb{Z}^n)$ is dense in $\mathbb{R}$. This conjecture was settled by Margulis [Ma2], [Ma3]. His proof is based on the study of the orbit structure of the action of $\mathrm{SO}(Q)$, the orthogonal group of Q, on the homogeneous space $\mathrm{SL}(n, \mathbb{R})/\mathrm{SL}(n, \mathbb{Z})$. Let us briefly indicate how these facts are related.

Set $H = \mathrm{SO}(Q)$, $G = \mathrm{SL}(n, \mathbb{R})$, $\Gamma = \mathrm{SL}(n, \mathbb{Z})$ and $\Omega = G/\Gamma$. The fact that Q is irrational implies that the orbit Hx_0 of $x_0 = \Gamma$ is not closed in Ω (see Proposition 2.4 below). Suppose now that – as in Hedlund's Minimality Theorem – we could prove that every non-closed H-orbit in Ω is dense. It follows then that $H\Gamma$ is dense in G and, hence,

$$\overline{Q(\mathbb{Z}^n)} = \overline{Q(H\Gamma\mathbb{Z}^n)} = Q(G\mathbb{Z}^n) = \mathbb{R}.$$

That indeed all non closed H-orbits in Ω are dense was first proved in [DM1], Theorem 2. We shall not give a proof for this fact. Instead, following the exposition in [DM3], we shall restrict ourselves to the arguments which are sufficient to show the density of $Q(\mathbb{Z}^n)$.

The density result of the orbits of H in Ω is now a special case of the spectacular work of Ratner on the so–called Raghunathan's conjecture. At the end of this chapter, we discuss related conjectures formulated by Dani and Margulis and Ratner's results solving all these conjectures.

§1 Oppenheim's Conjecture

Recall that a quadratic form Q on $\mathbb{R}^n$ is a function of the form

$$Q(x_1,\ldots,x_n) = \sum_{i,j=1}^n a_{ij}x_i x_j = x^t \sigma x, \qquad x^t = (x_1,\ldots,x_n) \in \mathbb{R}^n,$$

for a symmetric real matrix $\sigma = (a_{ij})_{1 \le i,j \le n}$. A quadratic form Q is indefinite if there exists $x \in \mathbb{R}^n \setminus \{0\}$ such that $Q(x) = 0$.

Let Q be a non-degenerate indefinite irrational quadratic form in n variables, $n \ge 3$.

Oppenheim's Conjecture (weak form) [Op1] *For every* $\varepsilon > 0$, *there exists a vector* $(z_1, z_2, \ldots, z_n) \in \mathbb{Z}^n \setminus \{0\}$ *such that*

$$|Q(z_1, z_2, \ldots, z_n)| < \varepsilon.$$

The stronger form of this conjecture, also due to Oppenheim [Op2], is obtained when the inequality $|Q(z_1, z_2, \ldots, z_n)| < \varepsilon$ is replaced by

$$0 < |Q(z_1, z_2, \ldots, z_n)| < \varepsilon.$$

As shown in [Op2], this modified conjecture implies that $Q(\mathbb{Z}^n)$ is dense in $\mathbb{R}$. This chapter is devoted to the proof of the following strengthening of Oppenheim's conjecture, first proved in [DM1].

1.1 Theorem. *Let* Q *be a non-degenerate indefinite irrational quadratic form in* n *variables,* $n \ge 3$. *Let* $\mathcal{P}$ *be the set of all primitive vectors in* $\mathbb{Z}^n$. *Then* $Q(\mathcal{P})$ *is dense in* $\mathbb{R}$.

Recall that a vector $p = (p_1, \ldots, p_n) \in \mathbb{Z}^n$ is *primitive* if the g.c.d. of $p_1, \ldots, p_n$ is 1.

1.2 Remark. The restriction $n \ge 3$ in Oppenheim's conjecture is necessary, as the following counterexample shows. Let a be a real algebraic number of degree 2. It is well known that there exists a constant $C > 0$ such that

$$\left| a - \frac{p}{q} \right| > \frac{C}{q^2}, \qquad \forall (p,q) \in \mathbb{Z}^2, q \ne 0. \tag{$*$}$$

Indeed, let $f(x) = a_0 x^2 + a_1 x + a_2$ be a polynomial of degree 2 with integer coefficients such that $f(a) = 0$. Then,

$$\left| f\left(\frac{p}{q}\right) \right| = \frac{1}{q^2} |a_0 p^2 + a_1 pq + a_2 q^2| \ge \frac{1}{q^2},$$

as $a_0 p^2 + a_1 pq + a_2 q^2$ is a non-zero integer. Clearly, it suffices to show $(*)$ for all $\frac{p}{q}$ in a compact interval around a. Then $|f'(x)|$ is bounded from above by some constant $m > 0$ on this interval. By the mean value theorem and the above inequality,

$$\left|a - \frac{p}{q}\right| \geq \frac{1}{m}|f(a) - f(p/q)| \geq \frac{1}{mq^2},$$

and this proves the claim.

Consider now the quadratic form on 2 variables $Q(x,y) = y^2 - a^2 x^2$. By the above inequality $(*)$,

$$|Q(x,y)| = \left|x^2(\frac{y}{x} - a)(\frac{y}{x} + a)\right| \geq C|a|$$

for all $x, y \in \mathbb{N}$. This implies that

$$\inf\{|Q(x,y)| \,|\, (x,y) \in \mathbb{Z}^2 \setminus \{(0,0)\}\} > 0.$$

During the proof of the above theorem, we shall have to use at several places Margulis' Lemma from the previous chapter (see Chap. V, Theorems 5.1 and 5.12).

§2 Proof of the Theorem – Preliminaries

Reduction to the case $n = 3$

We show that it is enough to consider the case $n = 3$. Using induction on n, this follows immediately from the next lemma.

2.1 Lemma. *If Q is a non-degenerate indefinite non-rational quadratic form in n variables, $n \geq 3$, there is a rational hyperplane L such that the restriction of Q to L is non-degenerate, indefinite and irrational.*

Proof Using orthogonal decomposition (see [La], Chap. XIV, §3, Theorem 1), we find linearly independent vectors $b_1, b_2, \ldots, b_{n-1} \in \mathbb{R}^n$ such that, denoting by L the linear span of $b_1, b_2, \ldots, b_{n-1}$ and by L' that of $b_1, b_2, \ldots, b_{n-2}$, the restriction $Q|_{L'}$ of Q to L' is non-degenerate and $Q|_L$ is indefinite and non-degenerate. By continuity, we may even assume that $b_1, \ldots, b_{n-1} \in \mathbb{Q}^n$. Indeed, let

$$A = (Q(b_i, b_j))_{1 \leq i,j \leq n-1}$$

be the matrix of $Q|_L$ with respect to $b_1, \ldots, b_{n-1}$. If $b'_1, \ldots, b'_{n-1}$ are close to $b_1, \ldots, b_{n-1}$, then the matrix $A' = (Q(b'_i, b'_j))_{i,j}$ of the restriction of Q to the linear span of $b'_1, \ldots, b'_{n-1}$ is close to A. Hence, the eigenvalues of A' are close to those of A. So, if $b'_1, \ldots, b'_{n-1}$ are close enough to $b_1, \ldots, b_{n-1}$, then A' is non-degenerate and A' has the same number of positive eigenvalues as A. Hence, the restriction of Q to the linear span of $b'_1, \ldots, b'_{n-1}$ is indefinite and non-degenerate and its restriction to the linear span of $b'_1, \ldots, b'_{n-2}$ is non-degenerate.

Now take an $x \in \mathbb{Q}^n \setminus L$ and define

$$L_t = \mathbb{R} - \mathrm{span}(b_1, b_2, \ldots, b_{n-1} + tx)$$

for $t \in \mathbb{Q}$ sufficiently small, so that $Q|_{L_t}$ is non-degenerate and indefinite. We claim that we may choose t such that, in addition, $Q|_{L_t}$ is not proportional to a rational quadratic form.

Assume, by contradiction, that this is not true. Then we find, for each t, an $\alpha_t \in \mathbb{R}$ such that

$$b_i^t \sigma b_j \in \alpha_t \mathbb{Q}, \quad 1 \le i, j \le n-2 \qquad\qquad (*)$$

$$b_i^t \sigma(b_{n-1} + tx) = b_i^t \sigma b_{n-1} + t b_i^t \sigma x \in \alpha_t \mathbb{Q}, \quad 1 \le i \le n-2 \qquad (**)$$

$$(b_{n-1} + tx)^t \sigma (b_{n-1} + tx) =$$
$$b_{n-1}^t \sigma b_{n-1} + 2t b_{n-1}^t \sigma x + t^2 x^t \sigma x \in \alpha_t \mathbb{Q}, \qquad\qquad (***)$$

where σ is the symmetric matrix corresponding to Q. Since $Q|_{L'}$ is non-zero, $(*)$ shows $\alpha_t \mathbb{Q} = \alpha_0 \mathbb{Q}$ for all t. Hence, $b_i^t \sigma b_{n-1} \in \alpha_0 \mathbb{Q}$ for $1 \le i \le n-1$ and $b_i^t \sigma x \in \alpha_0 \mathbb{Q}$ for $1 \le i \le n-2$, by taking $t = 0$ in $(**)$ and $(***)$. It follows, by $(***)$, that

$$2t b_{n-1}^t \sigma x + t^2 x^t \sigma x \in \alpha_0 \mathbb{Q}.$$

So, $2 b_{n-1}^t \sigma x + t x^t \sigma x \in \alpha_0 \mathbb{Q}$, and taking $t = 0$ yields

$$b_{n-1}^t \sigma x \in \alpha_0 \mathbb{Q} \quad \text{and} \quad x^t \sigma x \in \alpha_0 \mathbb{Q}.$$

Since $\{b_1, b_2, \ldots, b_{n-1}, x\}$ is a rational basis of the whole space, this shows that Q is proportional to a rational quadratic form, a contradiction. $\qquad\square$

So, we are reduced to quadratic forms on $\mathbb{R}^3$. Throughout this chapter, Q_0 will be the quadratic form on $\mathbb{R}^3$ defined by

$$Q_0(x_1, x_2, x_3) := 2x_1 x_3 - x_2^2. \qquad\qquad (1)$$

The corresponding matrix is

$$\sigma_0 = \begin{pmatrix} 0 & 0 & 1 \\ 0 & -1 & 0 \\ 1 & 0 & 0 \end{pmatrix},$$

and Q_0 is an indefinite form with signature $(2, 1)$. The general linear group $\mathrm{GL}(3, \mathbb{R})$ acts on the space of all quadratic forms Q on $\mathbb{R}^3$ by

$$gQ(p) := Q(g^{-1}p), \quad \forall p \in \mathbb{R}^3, \, g \in \mathrm{GL}(3, \mathbb{R})$$

the corresponding action on the space of all real symmetric 3×3 matrices A being

$$g \cdot A = (g^{-1})^t A g^{-1}.$$

By elementary linear algebra, $GL(3, \mathbb{R})$ has two orbits in the space $\Psi_{2,1}$ of all non-degenerate indefinite quadratic forms on $\mathbb{R}^3$. Indeed, any $Q \in \Psi_{2,1}$ may be transformed to Q_0 or to $-Q_0$, according to the sign of the determinant of the associated symmetric matrix σ_Q.

To prove the Oppenheim conjecture, it obviously suffices to consider the orbit $\Psi_{2,1}^+$ of the non-degenerate indefinite quadratic forms with positive determinant. We have a natural identification

$$GL(3, \mathbb{R})/O(Q_0) \longrightarrow \Psi_{2,1}^+, \quad gO(Q_0) \mapsto gQ_0,$$

where

$$O(Q_0) = \{g \in GL(3, \mathbb{R}) \,|\, gQ_0 = Q_0\} = \{g \in GL(3, \mathbb{R}) \,|\, g^t \sigma g = \sigma\} \cong O(2, 1)$$

is the orthogonal group of Q_0.

Let us say that two quadratic forms $Q, Q' \in \Psi_{2,1}^+$ are *equivalent* if there is a $g \in SL(3, \mathbb{Z})$ and a $c > 0$ such that

$$gQ = c^2 Q'.$$

Clearly, two equivalent quadratic forms have the same set of values on $\mathbb{Z}^3$, up to a constant factor. Any $Q \in \Psi_{2,1}^+$ is equivalent to gQ_0 for some $g \in SL(3, \mathbb{R})$. Hence, it suffices to prove Oppenheim's conjecture for the quadratic forms

$$gQ_0, \quad g \in SL(3, \mathbb{R}),$$

where $Q_0(x_1, x_2, x_3) = 2x_1 x_3 - x_2^2$.

Throughout this chapter, we set

$$G := SL(3, \mathbb{R}), \quad \Gamma := SL(3, \mathbb{Z})$$

and $H := SO(Q_0)$. (Observe that H is not connected as it acts transitively on the non-connected subset $\{x \in \mathbb{R}^3 \,|\, Q_0(x) + 1 = 0\}$ of $\mathbb{R}^3$.) We also set

$$\Omega := SL(3, \mathbb{R})/SL(3, \mathbb{Z}).$$

2.2 Lemma.

(i) *The connected component H_0 of the identity in H is generated by unipotent one-parameter subgroups.*

(ii) *The centralizer of H_0 in $GL(3, \mathbb{R})$ consists of multiples of the identity. (In particular, H_0 acts irreducibly on $\mathbb{R}^3$.)*

Proof The Lie algebra of H is

$$\mathfrak{h} = \left\{ X_{a,b,c} := \begin{pmatrix} a & b & 0 \\ c & 0 & b \\ 0 & c & -a \end{pmatrix} \,\Big|\, a, b, c \in \mathbb{R} \right\}$$

(see next exercise). Since $[X_{0,1,0}, X_{0,0,1}] = X_{1,0,0}$, it is clear that $\mathfrak{h}$ is generated by nilpotent endomorphisms. This proves (i) .

As to (ii), let $A \in M(3,\mathbb{R})$ commuting with all matrices from H. Then A commutes with all matrices from $\mathfrak{h}$. In particular, A commutes with the diagonal matrix $X_{1,0,0}$ with the diagonal entries $1, 0, -1$. Hence, A has to be a diagonal matrix. As A also commutes with $X_{0,1,0}$, it follows that A is a scalar matrix. $\square$

2.3 Exercise. Verify that the Lie algebra of H consists of the matrices $X_{a,b,c}$ introduced in the proof of the previous lemma.

Let $Q = g_Q^{-1} Q_0$, $g_Q \in G$, be an indefinite form in $\Psi_{2,1}^+$. The orthogonal group of Q is $SO(Q) = g_Q^{-1} H g_Q$. Let $x = g_Q \Gamma \in \Omega = G/\Gamma$.

2.4 Proposition. *If the form Q is irrational, then the H-orbit of x in $\Omega = G/\Gamma$ is not closed.*

Proof Suppose that Hx is closed. We have to show that Q is rational.

Let $x_0 = \Gamma \in \Omega$. Then the $SO(Q)$-orbit $SO(Q)x_0 = g_Q^{-1} Hx$ of x_0 is closed.

Let $\Delta := \Gamma \cap SO(Q)$. We claim that, if Q' is a quadratic form such that

$$\Delta \subseteq SO(Q'), \qquad\qquad (2)$$

then Q' is proportional to Q.

We first show that (2) implies that all unipotent one-parameter subgroups of $SO(Q)$ are contained in $SO(Q')$.

Fix $p \in \mathbb{R}^3$, and consider the function

$$f_p : SO(Q) \to \mathbb{R}, \quad g \mapsto Q'(g^{-1}p).$$

By (2), f_p factorizes through Δ to a continuous function

$$\tilde{f}_p : SO(Q)/\Delta \to \mathbb{R}.$$

Take a unipotent one-parameter group $\{u(t)\}_{t \in \mathbb{R}}$ in $SO(Q)$. The function

$$q : \mathbb{R} \to \mathbb{R}, \quad t \mapsto f_p(u(t))$$

is polynomial in t with values in $\tilde{f}_p(SO(Q)/\Delta) = f_p(SO(Q))$. By Margulis' Lemma (see Chap. V, Theorem 5.1), there exists a compact $K \subseteq \Omega$ such that the set

$$\{t \geq 0 | \ u(t)x_0 \in K\}$$

is unbounded.

Now, since $SO(Q)x_0$ is closed, the canonical mapping

$$\varphi : SO(Q)/\Delta \to SO(Q)x_0, \quad g\Delta \mapsto gx_0$$

is a homeomorphism. Hence, $K' = \varphi^{-1}(K)$ is compact. Therefore, $\tilde{f}_p(K')$ is a compact subset of $\mathbb{R}$. On the other hand, by the choice of K,

$$\{t \in \mathbb{R} \mid q(t) \in \tilde{f}_p(K')\}$$

is unbounded. Hence, the polynomial q is constant, that is, $Q'(u(t)p) = Q'(p)$ for all $t \in \mathbb{R}$. Since p was arbitrary, $\{u(t)\}_{t \in \mathbb{R}}$ is contained in $SO(Q')$, and the claim is proved.

Now, by the previous lemma, the connected component $SO_0(Q)$ of the identity in $SO(Q)$ is generated by unipotent one-parameter subgroups. Hence, $SO_0(Q) \subseteq SO(Q')$.

Let σ, σ' be the symmetric matrices corresponding to Q, Q', respectively. Then, for all $h \in SO_0(Q)$,

$$h\sigma'\sigma^{-1}h^{-1} = h\sigma'h^t \cdot (h^{-1})^t\sigma^{-1}h^{-1} = \sigma'\sigma^{-1}.$$

Now, by the previous lemma, the centralizer of H_0 in $GL(3,\mathbb{R})$ consists of scalars. This holds also for $SO_0(Q)$ which is conjugate to H_0. Hence, σ and σ' are proportional.

What we have shown so far is that the one-dimensional space $\mathbb{R}\sigma$ is the set of all symmetric real 3×3 matrices S such that

$$\gamma^t S\gamma = S, \quad \forall \gamma \in \Delta.$$

This is a system of linear equations where the variables are the entries of S and the coefficients are the entries of the γ's. Since there is a non-trivial solution, namely σ, and since the coefficients are integers, there is a non-trivial solution S with rational coefficients. The corresponding form is proportional to Q. Hence, Q is rational. $\qquad \square$

2.5 Remark. The previous lemma has the following converse. If Q is a rational quadratic form, then the $SO(Q)$-orbit of $x_0 = SL(3,\mathbb{Z})$ is closed. Moreover, $SO(Q)x_0$ is compact, if Q is rational and does not represent 0 non-trivially over $\mathbb{Q}$. These are special cases of results by A. Borel and Harish-Chandra (see [Bor], §8). We shall not use these facts.

Some subgroups of $SL(3,\mathbb{R})$

We shall have to consider the following subgroups of $H = SO(Q_0)$. Let

$$V_1 := \left\{ v_1(t) := \begin{pmatrix} 1 & t & \frac{t^2}{2} \\ 0 & 1 & t \\ 0 & 0 & 1 \end{pmatrix} \,\middle|\, t \in \mathbb{R} \right\}$$

and

$$D := \left\{ d(t) := \begin{pmatrix} a & 0 & 0 \\ 0 & 1 & 0 \\ 0 & 0 & a^{-1} \end{pmatrix} \,\middle|\, a > 0 \right\}.$$

Observe that V_1 is a unipotent one-parameter subgroup of H, that D normalizes V_1 and that

$$DV_1 = \left\{ \begin{pmatrix} a & at & \frac{at^2}{2} \\ 0 & 1 & t \\ 0 & 0 & a^{-1} \end{pmatrix} \,\middle|\, t \in \mathbb{R}, a > 0 \right\}$$

is a closed subgroup of H.

2.6 Lemma. *DV_1 is a cocompact subgroup of H.*

Proof Consider the natural action of H on the projective plane $\mathbb{PR}^2$. Let

$$\widetilde{C} := \{ p \in \mathbb{R}^3 \setminus \{0\} | \, Q_0(p) = 0 \} = \{ (p_1, p_2, p_3) \in \mathbb{R}^3 \setminus \{0\} | \, 2p_1 p_3 - p_2^2 = 0 \},$$

and $C = \{ \bar{p} | \, p \in \widetilde{C} \}$, where $\bar{p} \in \mathbb{PR}^2$ denotes the image of $p \in \mathbb{R}^3 \setminus \{0\}$. Then C is a compact subset of $\mathbb{PR}^2$ (since $\widetilde{C}$ is closed). It is clear C is invariant under H. Moreover, H acts transitively on C. Indeed, every $\bar{p} \in C$ is of the form $\bar{p} = \overline{(1, t, t^2/2)}$ for some $t \in \mathbb{R}$ or $\bar{p} = \overline{e_3}$, where $\{e_1, e_2, e_3\}$ is the standard basis of $\mathbb{R}^3$. Hence, $v_1(t)^t(\overline{e_1}) = \bar{p}$ in the first case and $\sigma_0(\overline{e_1}) = \bar{p}$ in the second case, where σ_0 is the matrix corresponding to the quadratic form Q_0 (observe that $v_1(t)^t$ and σ_0 are in H).

Let $R := \{ h \in H | \, h\overline{e_1} = \overline{e_1} \}$. Then H/R is homeomorphic to C and, hence, compact. Now, one has

$$R = \left\{ \begin{pmatrix} a & at & \frac{at^2}{2} \\ 0 & 1 & t \\ 0 & 0 & a^{-1} \end{pmatrix} \,\middle|\, t \in \mathbb{R}, a \in \mathbb{R}, a \neq 0 \right\}. \tag{3}$$

Indeed, if $h \in R$, then the first column of h is ae_1 for some $a \in \mathbb{R}$, and a computation using the fact that

$$h^t \begin{pmatrix} 0 & 0 & 1 \\ 0 & -1 & 0 \\ 1 & 0 & 0 \end{pmatrix} h = \begin{pmatrix} 0 & 0 & 1 \\ 0 & -1 & 0 \\ 1 & 0 & 0 \end{pmatrix}$$

and that $\det h = 1$ proves the claim. Hence, DV_1 is a subgroup of index two in R. Since R is cocompact in H, this finishes the proof. □

2.7 Exercise. Verify the above claim (3).

We shall need some further subgroups of $SL(3, \mathbb{R})$ which are *not* contained in H. Let

$$V_2 := \left\{ v_2(s) = \begin{pmatrix} 1 & 0 & s \\ 0 & 1 & 0 \\ 0 & 0 & 1 \end{pmatrix} \middle| s \in \mathbb{R} \right\}$$

$$V := V_1 \cdot V_2 = \left\{ \begin{pmatrix} 1 & t & s \\ 0 & 1 & t \\ 0 & 0 & 1 \end{pmatrix} \middle| s, t \in \mathbb{R} \right\}.$$

Observe that V, V_1 and V_2 are abelian subgroups, all normalized by D, and that V_2 centralizes V_1.

For later use, we record some facts about the discrete subgroups in DV.

2.8 Lemma. *Let Δ be a discrete subgroup of $DV = DV_1V_2$. Then*

(i) *either $\Delta \subseteq V$ or Δ is a cyclic group generated by an element vdv^{-1} for some $d \in D \setminus \{e\}, v \in V$. In the latter case, if $\Delta \subseteq DV_1$, then $v \in V_1$;*

(ii) *if Δ is generated by an element $vdv^{-1}, d \in D \setminus \{e\}, v \in V$, the projection of Δ on the V_2-component is discrete.*

Proof

(i) Assume that Δ is not contained in V. Let

$$d := \begin{pmatrix} \delta & 0 & 0 \\ 0 & 1 & 0 \\ 0 & 0 & \delta^{-1} \end{pmatrix} \in D \setminus \{e\}$$

$$w := \begin{pmatrix} 1 & s & t \\ 0 & 1 & s \\ 0 & 0 & 1 \end{pmatrix} \in V$$

be such that $dw \in \Delta$. Then

$$dw = \begin{pmatrix} \delta & \delta s & \delta t \\ 0 & 1 & s \\ 0 & 0 & \delta^{-1} \end{pmatrix}$$

has an eigenbasis consisting of the vectors

$$v^1 = \begin{pmatrix} 1 \\ 0 \\ 0 \end{pmatrix}, v^2 = \begin{pmatrix} \frac{\delta s}{1 - \delta^{-1}} \\ 1 \\ 0 \end{pmatrix}, v^3 = \begin{pmatrix} \frac{(\delta s)^2 + \delta t}{\delta^{-1} - \delta} \\ \frac{\delta s}{1 - \delta^{-1}} \\ 1 \end{pmatrix}.$$

So, for the matrix $v \in V$ with columns v^1, v^2, v^3, we have $dw = vdv^{-1}$. Hence,

$$\Delta' := v^{-1} \Delta v$$

is a discrete subgroup containing d. For $a \in D$, $u \in V$ with $au \in \Delta'$, one has

$$d^j a u d^{-j} = a d^j u d^{-j} \in \Delta' \quad \forall j \in \mathbb{Z}. \tag{4}$$

On the other hand, $d^j u d^{-j} \to e$ for $j \to \infty$ or $j \to -\infty$, so that

$$d^j a u d^{-j} \to a \in \Delta',$$

by (4). Since Δ' is discrete, this shows that $d^j a u d^{-j} = a$ for large j and hence $u = e$. So, Δ' is a discrete subgroup of the group D. The latter is topologically isomorphic to the reals. Therefore Δ' and, hence, Δ are cyclic.

(ii) Write $v = v_1(s)v_2(t) \in V_1 V_2$ and $d = \begin{pmatrix} \delta & 0 & 0 \\ 0 & 1 & 0 \\ 0 & 0 & \delta^{-1} \end{pmatrix}$. Then, for j in

$\mathbb{Z}$, one has

$$
\begin{aligned}
v d^j v^{-1} &= d^j (d^{-j} v_1(s) d^j)(d^{-j} v_2(t) d^j) v_1(-s) v_2(-t) \\
&= d^j v_1(\delta^{-j} s - s) v_2(\delta^{-2j} t - t).
\end{aligned}
$$

Clearly, 0 is isolated in the set

$$\{\delta^{-2j} t - t \mid j \in \mathbb{Z}\} = \{(\delta^{-2j} - 1)t \mid j \in \mathbb{Z}\}.$$

This proves the claim. $\qquad\qquad\qquad\qquad\qquad\qquad\qquad\qquad\qquad\quad\square$

Let Q_1 be the (degenerate) form

$$Q_1(x_1, x_2, x_3) := x_3^2.$$

Observe that

$$Q_0(v_2(t)p) = Q_0(p) + 2tQ_1(p), \quad \forall t \in \mathbb{R}, \ p \in \mathbb{R}^3. \tag{5}$$

We shall also need later the following fact.

2.9 Lemma. *Let $h \in H$ and let $v = v_2(t) \in V_2 \setminus \{e\}$ be such that vh lies in HV_2. Then h is an upper triangular matrix. Moreover, if all the diagonal entries of h are positive, then $h \in DV_1$.*

Proof Let $h' \in H$ and $s \in \mathbb{R}$ be such that $v_2(t)h = h'v_2(s)$. Then, by (5), for all $p \in \mathbb{R}^3$

$$Q_0(v_2(t)hp) = Q_0(hp) + 2tQ_1(hp) \qquad \text{and}$$
$$Q_0(h'v_2(s)p) = Q_0(v_2(s)p) = Q_0(p) + 2sQ_1(p).$$

Hence,

$$Q_1(hp) = \frac{s}{t} Q_1(p).$$

Therefore, h leaves the radical of the form Q_1 invariant. This radical is $\mathbb{R}e_1 \oplus \mathbb{R}e_2$. Hence, h is of the form

$$h = \begin{pmatrix} p_1 & q_1 & * \\ p_2 & q_2 & * \\ 0 & 0 & * \end{pmatrix}.$$

Then $-p_2^2 = Q_0(he_1) = Q_0(e_1) = 0$. Hence, h is upper triangular.

The last assertion follows from the last step in the proof of Lemma 2.6. $\square$

Let

$$V_2^+ := \{v_2(s) \mid s > 0\} \quad \text{and} \quad V_2^- := \{v_2(s) \mid s < 0\}.$$

The strategy in proving Theorem 1.1 is to show that, if Q is irrational, then the closure of $Hg_Q\Gamma$ in Ω contains a V_2^+ or a V_2^--orbit. Once this is done, to finish the proof, we only have to apply the next lemma. Recall that $\mathcal{P}$ denotes the primitive integral vectors in $\mathbb{R}^3$.

2.10 Lemma. *Let $g \in G$. Then*

$$Q_0(V_2^+ g\mathcal{P}) = Q_0(V_2^- g\mathcal{P}) = \mathbb{R}.$$

In particular, if the form $Q = g_B^{-1}Q_0$ is such that $\overline{Hg_Q\Gamma}$ contains a V_2^+ or a V_2^--orbit, then $Q(\mathcal{P})$ is dense in $\mathbb{R}$.

Proof Let $s \in \mathbb{R}$, and define $s_0 := \min\{s, 0\}$. The set

$$C := \{p \in \mathbb{R}^3 \mid Q_0(gp) < s_0 \quad \text{and} \quad Q_1(gp) > 0\}.$$

is a non-empty open subset, invariant under multiplication with any real number $t \geq 1$. Therefore, by the following lemma (Lemma 2.11), it contains a primitive integer vector p_0. Now set

$$t := \frac{s - Q_0(gp_0)}{2Q_1(gp_0)} > 0.$$

Then, by (5),

$$Q_0(v_2(t)gp_0) = Q_0(gp_0) + 2tQ_1(gp_0) = s.$$

Since s was arbitrary, this settles the first statement.

As to the second statement, choose $g \in G$ with $V_2^+ g\Gamma \subseteq Hg_Q\Gamma$. Observe that $\Gamma\mathcal{P} = \mathcal{P}$. Hence, by continuity,

$$\overline{Q(\mathcal{P})} = \overline{Q_0(Hg_Q\Gamma\mathcal{P})}$$

contains $Q_0(\overline{Hg_Q\Gamma\mathcal{P}})$. Therefore, $\overline{Q(\mathcal{P})}$ contains

$$Q_0(V_2^+ g\Gamma\mathcal{P}) = Q_0(V_2^+ g\mathbb{Z}^3) = \mathbb{R}.$$

The proof for V_2^- is similar. $\square$

2.11 Lemma. *Let C be a non-empty open set in $\mathbb{R}^n$, where $n \geq 2$, such that $tC \subseteq C$ for all $t \geq 1$. Then C contains a primitive vector in $\mathbb{Z}^n$.*

Proof As C is open and $n \geq 2$, there exist linearly independent rational vectors $p, q \in C$ such that $tp + (1 - t)q \in C$ for all $0 \leq t \leq 1$. Replacing p, q by multiples kp, kq for some positive integer k, we may assume that $p, q \in \mathbb{Z}^n$. Since $tC \subseteq C$ for all $t \geq 1$, we have

$$sp + tq \in C \qquad \forall s, t \geq 0, \ s + t \geq 1. \tag{6}$$

Let m be the g.c.d. of the coordinates of p. There exists $\gamma \in \mathrm{SL}(n, \mathbb{Z})$ such that $\gamma p = m e_1$. (It is clear that $\mathrm{SL}(n, \mathbb{Z})$ acts transitively on the set of all integer vectors with the same g.c.d.)

Let $m_1, \ldots, m_n$ be the coordinates of γq. Then $m_i \neq 0$ for some $i \geq 2$, since γp and γq are linearly independent. Let p_0 a positive prime number such that $p_0 \geq m_1$ and such that p_0 does not divide m_i.

Let

$$r = (p_0 - m_1)e_1 + \gamma q = (p_0, m_2, \ldots, m_n).$$

Since p_0 does not divide m_i, r is a primitive vector. Hence, the same is true for

$$\gamma^{-1} r = \gamma^{-1}(p_0 - m_1)e_1 + q = \frac{1}{m}(p_0 - m_1)p + q.$$

Since $p_0 - m_1 \geq 0$, it follows from (6) that $\gamma^{-1} r \in C$. $\square$

§3 **Existence of Minimal Closed Subsets**

Let $Q = g_Q^{-1} Q_0$ be irrational, and $x = g_Q \Gamma \in \Omega$. In order to show that $\overline{Hx}$ contains a V_2^+ or a V_2^--orbit, a crucial step is to find non-empty closed subsets X_1, Y with

$$\overline{Hx} \supset X_1 \supset Y,$$

which are minimal with respect to the action of DV_1 and V_1, respectively (compare with Chap. IV, §2).

Now, the existence of such sets is clear if $\overline{Hx}$ is compact (see Chap. IV, Lemma 2.1). We are going to show that minimal closed DV_1 or V_1-invariant sets still exist in the general case. However, the proof is considerably more difficult and requires the uniform version of Margulis' Lemma given in Chap. V.

3.1 Lemma. *For $\lambda \in \mathbb{R}$, let $a(\lambda)$ denote the diagonal matrix with diagonal entries $\lambda, 1$, and λ^{-1}. There is a compact set $K \subseteq \Omega = G/\Gamma$ with the following property. For every $x = g\Gamma \in \Omega$, there exists $\lambda_0 > 0$ such that the set*

$$A_\lambda := \{t \geq 0 | \ v_1(t)a(\lambda)x \in K\}$$

is unbounded for all $\lambda \geq \lambda_0$.

Proof By the uniform version of Margulis' Lemma (Chap. V, Theorem 5.17), applied to $\sigma = 1$, there exists a compact set $K \subseteq \Omega$ such that, for each $x = g\Gamma \in \Omega$ and for any unipotent one-parameter subgroup $\{u_t\}_{t \in \mathbb{R}}$, either the set $\{t \geq 0 | u_t x \in K\}$ is unbounded or there exists a non-zero subgroup Δ of $\mathbb{Z}^3$ such that $\Delta_{\mathbb{R}}$ is invariant under $\{g^{-1} u_t g\}_{t \in \mathbb{R}}$ and such that $d(u_t g \Delta) = d(g\Delta) < 1$ for all $t \in \mathbb{R}$. (Recall that $d(\Delta)$ is the determinant of Δ; see Chap. V, §5.)

Fix $g \in G$. Let $x = g\Gamma$, and let $W_1 = \mathbb{R}e_1$ and $W_2 = \mathbb{R}e_1 \oplus \mathbb{R}e_2$. Define

$$\lambda_0 := \max\left(\frac{1}{d(g\mathbb{Z}^3 \cap W_1)}, \frac{1}{d(g\mathbb{Z}^3 \cap W_2)}, 1\right).$$

Let $\lambda \geq \lambda_0$. Let Δ be a non-discrete subgroup of $\mathbb{Z}^3$ such that $\Delta_{\mathbb{R}}$ is a proper subspace, invariant under the unipotent one-parameter subgroup

$$g^{-1} a(\lambda)^{-1} V_1 a(\lambda) g = g^{-1} V_1 g.$$

Then $g\Delta_{\mathbb{R}}$ is a V_1-invariant proper subspace. Clearly, this implies that $g\Delta_{\mathbb{R}}$ is equal either to W_1 or to W_2. Both spaces W_1 and W_2 are invariant under $a(\lambda)$ and $\det(a(\lambda)|_{W_i}) = \lambda$ for $i = 1, 2$. Hence, $d(a(\lambda)g\Delta) = \lambda d(g\Delta)$. Moreover, since $g\Delta$ is contained in $g\mathbb{Z}^3 \cap W_1$ or in $g\mathbb{Z}^3 \cap W_2$, we have

$$d(g\Delta) \geq d(g\mathbb{Z}^3 \cap W_1) \quad \text{or} \quad d(g\Delta) \geq d(g\mathbb{Z}^3 \cap W_2).$$

Hence,

$$d(a(\lambda)g\Delta) \geq \frac{\lambda}{\lambda_0} \geq 1 = \sigma.$$

Therefore, by the choice of K, the set A_λ is unbounded for $\lambda \geq \lambda_0$. $\square$

3.2 Proposition. *Any closed non-empty DV_1-invariant subset of Ω contains a minimal non-empty closed DV_1-invariant set.*

Proof Let Z be a closed non-empty DV_1-invariant subset of Ω, and let $\{X_\alpha\}_\alpha$ be a totally ordered family of closed non-empty DV_1-invariant subsets of Z. Choose a compact subset K of Ω as in Lemma 3.1 above. Then each $X_\alpha \cap K$ is non-empty. Since K is compact, $\bigcap_\alpha (X_\alpha \cap K)$ is also non-empty. Hence, the closed DV_1-invariant subset $\bigcap_\alpha X_\alpha$ is non-empty. Apply now Zorn's lemma. $\square$

In order to apply Chap. IV, Lemma 2.3, we shall need to know that the DV_1-orbits are not closed.

3.3 Theorem. *There are no closed DV_1-orbits in Ω.*

Proof Assume, by contradiction, that $DV_1 x$ is closed for some $x \in \Omega$.

The subgroup

$$\Delta := \{g \in DV_1 |\ gx = x\}.$$

of DV_1 is discrete and, since the orbit is closed, the mapping

$$\Phi : DV_1/\Delta \longrightarrow DV_1 x,\ g\Delta \mapsto gx,$$

is a homeomorphism. By Lemma 2.8, Δ is either contained in V_1 or Δ is a cyclic group generated by some element vdv^{-1}, where $d \in D \setminus \{e\}$, $v \in V_1$. Choose a compact subset K of Ω as in Lemma 3.1 above.

Suppose $\Delta \subseteq V_1$. Let $\{\lambda_n\}_n$ be a sequence of positive real numbers with $\lambda_n \to \infty$. Since $\Delta \subseteq V_1$, for any sequence $\{t_n\}_n \in \mathbb{R}$, the sequence $\{a(\lambda_n)v_1(t_n)\Delta\}_{n\in\mathbb{N}}$ has no convergent subsequence in DV_1/Δ. Hence, $\{a(\lambda_n)v_1(t_n)x\}_{n\in\mathbb{N}}$ has no convergent subsequence, since Φ is a homeomorphism. Since (by the choice of K) for all large λ, there exists $t \geq 0$ such that

$$v_1(t)a(\lambda)x = a(\lambda)v_1(\lambda^{-1}t)x \in K,$$

this is a contradiction.

Suppose now that Δ is a cyclic group generated by vdv^{-1}, where $d \in D \setminus \{e\}$, $v \in V_1$. Then, for $\lambda > 0$, the set $V_1 a(\lambda)\Delta = a(\lambda)V_1\Delta$ is closed and the mapping

$$\mathbb{R} \longrightarrow V_1 a(\lambda)\Delta/\Delta,\ t \mapsto v_1(t)a(\lambda)\Delta$$

is a homeomorphim. But, since Φ is a homeomorphism, this implies that, for $\lambda > 0$, the mapping $t \mapsto v_1(t)a(\lambda)x$ is a homeomorphism from $\mathbb{R}$ onto the closed set $V_1 a(\lambda)x$. Now the intersection

$$\{v_1(t)a(\lambda)x |\ t \geq 0\} \cap K$$

is compact. On the other hand, for large λ,

$$\{t \geq 0|\ v_1(t)a(\lambda)x \in K\}$$

is unbounded, this is a contradiction. $\square$

We now turn to the existence of V_1-minimal sets. We first introduce some V_1-invariant closed subsets.

3.4 Lemma. *Let $\varrho > 0$ and consider, for $i = 1, 2$, the set*

$$A(i, \varrho) := \{g\Gamma \in \Omega |\ g\mathbb{Z}^3 \cap W_i \text{ spans } W_i \text{ and } d(g\mathbb{Z}^3 \cap W_i) = \varrho\},$$

where $W_1 = \mathbb{R}e_1$ and $W_2 = \mathbb{R}e_1 \oplus \mathbb{R}e_2$. Then $A(i, \varrho)$ is a closed V_1-invariant subset of Ω.

Proof Let $g\Gamma \in A(1, \varrho)$. Let $m \in \mathbb{Z}^3$ be such that gm is a generator of the group $g\mathbb{Z}^3 \cap W_1$. Then m is necessarily primitive. Hence, there exists $\gamma \in \Gamma$ with $\gamma e_1 = m$. Let a be the diagonal matrix with diagonal entries

$\rho^{-1}, 1, 1$. Then, clearly, $ag\gamma e_1 = e_1$. Hence,

$$A(1, \varrho) = G_1 a^{-1} \Gamma / \Gamma = a^{-1} G_1 \Gamma / \Gamma, \qquad \text{where} \qquad G_1 = \{g \in G|\ g e_1 = e_1\}.$$

Let $g\Gamma \in A(2, \varrho)$. Let $m_1, m_2 \in \mathbb{Z}^3$ be such that $g m_1, g m_2$ is a $\mathbb{Z}$-basis of $g\mathbb{Z}^3 \cap W_2$. There exists $\gamma \in \Gamma$ with $\gamma e_1 = m_1, \gamma e_2 = m_1$, since m_1, m_2 is a $\mathbb{Z}$-basis of the complete discrete subgroup $\mathbb{Z}^3 \cap g^{-1} W_2$ of $\mathbb{Z}^3$ (see next exercise). Let b be the diagonal matrix with diagonal entries $1, 1, \rho$. Then $bg\gamma e_1$ has e_3^t as last row. Hence,

$$A(2, \varrho) = G_2 b^{-1} \Gamma / \Gamma = b^{-1} G_2 \Gamma / \Gamma, \qquad \text{where} \qquad G_2 = \{g \in G|\ g^t e_3 = e_3\}.$$

It is now clear that $A(1, \varrho)$ and $A(2, \varrho)$ are V_1-invariant.

To show that $A(1, \varrho)$ is closed, observe that $\Gamma \cdot s e_1$ is a discrete subset of $\mathbb{R}^3$, for each real number s. Hence, if d is a diagonal matrix, then $d e_1 = s e_1$ for a suitable $s \in \mathbb{R}$, and $\Gamma \cdot d e_1$ is discrete and, hence, closed in G. Therefore, the set

$$\Gamma d G_1 = \{g \in G|\ g e_1 \in \Gamma \cdot d e_1\}$$

is closed for any diagonal matrix d. Hence, the same is true for $G_1 d \Gamma = (\Gamma d^{-1} G_1)^{-1}$ and for $A(1, \varrho) = G_1 a^{-1} \Gamma / \Gamma$, by definition of the quotient topology.

The fact that $A(2, \varrho)$ is closed is proved in a similar way, replacing the natural action by the transpose action of G on $\mathbb{R}^3$. □

3.5 Exercise. Let W be a linear subspace of $\mathbb{R}^n$, and let $m_1, \dots, m_r$ be a $\mathbb{Z}$-basis of the complete discrete subgroup $\mathbb{Z}^n \cap W$. Show that $m_1, \dots, m_r$ can be extended to a basis of $\mathbb{Z}^n$.

We shall also need the following general fact.

3.6 Lemma. *Let X be a locally compact space, and let $\{\varphi_t\}_{t \in \mathbb{R}}$ be a one-parameter group of homeomorphisms of X. Suppose that there exists a compact subset K of X such that, for each $x \in X$, the sets*

$$\{t \geq 0|\ \varphi_t x \in K\} \quad \text{and} \quad \{t \leq 0|\ \varphi_t x \in K\}$$

are unbounded. Then X is compact.

Proof Set $\varphi := \varphi_1$. Replacing, if necessary, K by the larger compact set

$$\{\varphi_s x|\ -1 \leq s \leq 1, x \in K\},$$

we may assume that, for each $x \in X$, the sets

$$\{j \in \mathbb{N}|\ \varphi^j x \in K\} \quad \text{and} \quad \{j \in \mathbb{N}|\ \varphi^{-j} x \in K\}$$

are unbounded.

Let K_1 be a compact neighbourhood of K. Observe that, by assumption,

$$X = \bigcup_{j=1}^{\infty} \varphi^{-j}(K_1).$$

We claim that there exists $N \in \mathbb{N}$ such that

$$X = \bigcup_{j=1}^{N} \varphi^{-j}(K_1).$$

Clearly, this would imply that X is compact. Assume that no such N exists. Then there exists, for each $n \in \mathbb{N}$, a point $x_n \in X$ such that

$$\varphi^j(x_n) \in X \setminus K_1, \qquad \forall\, 1 \le j \le n.$$

Now, by assumption, we also have

$$X = \bigcup_{j=1}^{\infty} \varphi^j(K_1).$$

For each $n \in \mathbb{N}$, let $j_n \in \mathbb{N}$ be such that

$$y_n = \varphi^{-j_n} x_n \in K_1 \quad \text{and} \quad \varphi^{-j} x_n \notin K_1, \quad \forall\, 0 \le j < j_n.$$

Then

$$\varphi^j y_n \in X \setminus K_1, \qquad \forall\, j \in \mathbb{N},\ n \ge j.$$

Since K_1 is compact, we may assume that $y = \lim_n y_n \in K_1$ exists. Then, for each $j \in \mathbb{N}$, $\varphi^j y = \lim_n \varphi^j y_n$ is in the closure of $X \setminus K_1$. But this closure does not meet K. Hence,

$$\varphi^j y \notin K, \qquad \forall\, j \in \mathbb{N}.$$

This is a contradiction. □

3.7 Proposition. *Any non-empty closed V_1-invariant set Z of Ω contains a compact non-empty V_1-invariant set. Hence, Z contains a non-empty V_1-minimal set and any such set is compact.*

Proof Of course, it suffices to prove the first assertion. For $i = 1, 2$, the sets $A(i, \varrho)_{\varrho > 0}$ as in Lemma 3.4 are V_1-invariant, closed and mutually disjoint. Replacing Z by a (possibly smaller) closed invariant set, we may assume that Z is either contained in some $A(i, \varrho)$ or that $Z \cap A(i, \varrho) = \emptyset$ for all sets $A(i, \varrho)$.

Fix $\sigma \le 1$ such that $\sigma \le \varrho$ if $Z \subseteq A(i, \varrho)$ for $i = 1$ or $i = 2$. Let $K \subseteq \Omega$ be a compact subset given by the uniform version of Margulis' Lemma (Theorem V.5.17), applied to σ. We claim that, for each $x \in Z$, the sets

$$\{t \ge 0|\ v_1(t)x \in K\} \quad \text{and} \quad \{t \le 0|\ v_1(t)x \in K\}$$

are unbounded. The previous lemma implies then that Z is compact, finishing the proof.

Assume, by contradiction, that there exists $x = g\Gamma \in Z$ such that one of the above sets is bounded. Then, by Margulis' Lemma applied to $\{v_1(t)\}_{t\in\mathbb{R}}$ or to $\{v_1(-t)\}_{t\in\mathbb{R}}$ and to $x = g\Gamma$, there exists a non-zero subgroup Δ of $\mathbb{Z}^3$ such that $\Delta_\mathbb{R}$ is invariant under $g^{-1}V_1g$ and that

$$d(v_1(t)g\Delta) = d(g\Delta) < \sigma.$$

Since $\sigma \leq 1$, $g\Delta_\mathbb{R}$ is a proper subspace, invariant under V_1. We conclude (in the same way as in the proof of Lemma 3.1 above) that $g\Delta_\mathbb{R}$ is equal either to W_1 or to W_2 (with W_i as in Lemma 3.4 above.) This implies that $x = g\Gamma \in Z\cap A(i,\varrho)$, where ϱ is the determinant of $g\mathbb{Z}^3\cap W_i$. Hence, $Z \subseteq A(i,\varrho)$, by our assumption on Z. Since

$$\varrho \leq d(g\Delta) < \sigma,$$

this is a contradiction to our choice of σ. Hence, the sets as above are unbounded and the proof is complete. $\qquad\qquad\square$

§4 Orbits of One-Parameter Groups of Unipotent Linear Transformations

Let M be a subset of $\mathbb{R}^2\setminus\{0\}$, invariant under the action of the group N of all unipotent upper triangular matrices of $SL(2,\mathbb{R})$. Assume that M does not meet the real axis, and that the closure $\overline{M}$ contains some point from the real axis. Then $\overline{M}$ contains the whole real axis. This elementary fact played an important rôle in the proof of Hedlund's Minimality Theorem (see end of Chap. IV, §2).

We shall need a generalization of this result to actions of one-parameter groups of linear transformations on a vector space of arbitrary dimension. Observe that, in the above example, the real axis is the fixed point set of N.

Let E be a finite-dimensional real vector space. Recall also that a mapping $f:\mathbb{R}\to E$ is *polynomial* if there exists a basis $\{e_1,e_2,\dots,e_n\}$ of E with

$$f(t) = \sum_{j=1}^{n} p_j(t)e_j$$

where all p_j's are polynomials (this definition is independent of the choice of the basis).

4.1 Lemma. *Let* $\{u(t)\}_{t \in \mathbb{R}}$ *be a unipotent one-parameter subgroup on a finite-dimensional real vector space* E, *and let*

$$L := \{x \in \mathbb{R}^n \,|\, u(t)x = x \; \forall t \in \mathbb{R}\}$$

be the space of the common fixed points of the $u(t)$ *'s. Let* M *be a subset of* $E \setminus L$. *Then, for every* $p_0 \in L \cap \overline{M}$, *there exists a nonconstant polynomial arc through* p_0 *in* $L \cap \overline{u(\mathbb{R})M}$. *More precisely, one finds a non-constant polynomial* $\varphi : \mathbb{R} \to L$ *and sequences* $\{t_i\}_i, \{m_i\}_i$ *in* $\mathbb{R}$ *and* M, *respectively, with*

$$\varphi(0) = p_0 \quad and \quad \lim_{i \to \infty} u(st_i)(m_i) = \varphi(s),$$

for all $s \in \mathbb{R}$.

Proof Let τ be the nilpotent endomorphism defining $\{u(t)\}_{t \in \mathbb{R}}$. With respect to a Jordan basis $\{e_j^{(k)} | 1 \leq k \leq \ell, 1 \leq j \leq r_k\}$, τ has a matrix of the form

$$\tau = \begin{pmatrix} J_1 & & & \\ & J_2 & & \\ & & \ddots & \\ & & & J_\ell \end{pmatrix}$$

and each Jordan block $J_k, 1 \leq k \leq \ell$ is of the form

$$J_k = \begin{pmatrix} 0 & 1 & & \\ & 0 & 1 & \\ & & \ddots & \ddots \\ & & & 0 & 1 \\ & & & & 0 \end{pmatrix}.$$

To simplify notation, we assume that there exists only one Jordan block and suppress therefore the index k. Then, for each $t \in \mathbb{R}$, we have

$$u(t) = \exp(tJ) = \begin{pmatrix} 1 & t & \frac{t^2}{2} & \cdots & \frac{t^{r-1}}{(r-1)!} \\ & 1 & t & \cdots & \frac{t^{r-2}}{(r-2)!} \\ & & \ddots & \ddots & \vdots \\ & & & 1 & t \\ & & & & 1 \end{pmatrix}$$

and $L = \mathbb{R}e_1$.

Set $\theta(m) := \max\{|m(j)|^{\frac{1}{j-1}} | 2 \leq j \leq r\}$ for $m = \sum_{j=1}^r m(j)e_j \in E$. So,

$$\left| \frac{m(j)}{\theta(m)^{j-1}} \right| \leq 1 \quad \forall j \geq 2 \quad and \quad \left| \frac{m(j_0)}{\theta(m)^{j_0-1}} \right| = 1 \quad \text{for some} \quad j_0 \geq 2. \qquad (*)$$

Let $\{m_i\}_i \subseteq M$ be a sequence with $m_i \to p_0$. By $(*)$, we may assume, upon passing to a subsequence, that

$$\lambda(j) := \lim_{i \to \infty} \frac{m_i(j)}{\theta(m_i)^{j-1}}$$

exists for each j and that, for some $j_0 \geq 2$, $|\lambda(j_0)| = 1$.

Define a polynomial mapping $\varphi : \mathbb{R} \to E$ by

$$\varphi(s) := \sum_{j=1}^{r} \frac{\lambda(j)}{(j-1)!} s^{j-1} e_1.$$

Since $|\lambda(j_0)| = 1$ and $j_0 \geq 2$, φ is not constant. Moreover, $\varphi(s) \in L$ for all $s \in \mathbb{R}$. Let $t_i := 1/\theta(m_i)$. Then, for all $s \in \mathbb{R}$,

$$u(st_i)(m_i) = \sum_{j=1}^{r} m_i(j) u(st_i) e_j$$

$$= \sum_{j=1}^{r} \sum_{\nu=0}^{j-1} m_i(j) t_i^{\nu} \frac{s^{\nu}}{\nu!} e_{j-\nu}$$

But

$$\lim_{i \to \infty} m_i(j) t_i^{\nu} = \lim_{i \to \infty} \frac{m_i(j)}{\theta(m_i)^{\nu}} = \begin{cases} 0 & \text{if } \nu < j - 1 \\ \lambda(j) & \text{if } \nu = j - 1 \end{cases}$$

since $\lim_{i \to \infty} m_i(j) = 0$ for $j \geq 2$. Hence, for all $s \in \mathbb{R}$,

$$u(st_i)(m_i) \to \varphi(s) \quad \text{and} \quad \varphi(0) = \lambda(1) e_1 = p_0,$$

as desired. $\qquad\qquad\qquad\qquad\qquad\qquad\qquad\qquad\qquad\qquad\qquad\qquad\qquad\square$

§5 Proof of the Theorem – Conclusion

We are now able to give the proof of Theorem 1.1.

Let $Q = g_Q^{-1} Q_0$ be an indefinite irrational form, and let $x = g_B \Gamma \in \Omega$. We know from Proposition 2.4, that the closure $\overline{Hx}$ of the H-orbit of x is not closed. To complete the proof, it suffices, by Lemma 2.10, to show that $\overline{Hx}$ contains a V_2^+ or a V_2^--orbit.

According to the Propositions 3.2 and 3.7, there exist a closed, non empty DV_1-minimal subset X_1 contained in $\overline{Hx}$ and a closed, non-empty V_1-minimal set Y contained in X_1. So,

$$\overline{Hx} \supset X_1 \supset Y.$$

Observe that Y is compact (see Proposition 3.7). Fix $y \in Y$. As in the proof of Hedlund's Minimality Theorem, we shall consider the following subset of G:

$$R := \{g \notin H \mid gy \in \overline{Hx}\}.$$

The following is a crucial fact.

5.1 Lemma. *The group unit e belongs to $\overline{R}$.*

Proof Assume, by contradiction, that $e \notin \overline{R}$. Then there is a neighbourhood U of e in G with $Uy \cap \overline{Hx} \subseteq Hy$. Hence, since $y \in \overline{Hx}$, $Uy \cap Hx$ is a non-empty open subset of $\overline{Hx}$ and

$$\emptyset \neq Uy \cap Hx \subseteq Hy \cap Hx.$$

Hence, $Hx = Hy$.

The set $\overline{X_1 \setminus Hy}$ is a DV_1–invariant closed subset of X_1. Moreover, $\overline{X_1 \setminus Hy}$ does not contain y (since $Uy \cap Hx \subseteq Hy$). Hence, $X_1 \subseteq Hy$, by minimality of X_1. Therefore, $HX_1 = Hy$, as $y \in X_1$.

Since, by Lemma 2.6, DV_1 is cocompact, there exists a compact subset $K \subseteq H$ such that $H = KDV_1$. Hence,

$$KX_1 = K(DV_1)X_1 = Hy.$$

On the other hand, KX_1 is closed, since K is compact. Hence, $Hy = Hx$ is closed. This is a contradiction.   □

Proof of Theorem 1.1   We are going to show that $\overline{Hx}$ contains either $V_2^+ y$ or $V_2^- y$. By Lemma 2.10, this will finish the proof of the theorem.

Let E be the vector space of the symmetric real 3×3 matrices. The group G acts on E via

$$(g, \sigma) \mapsto (g^{-1})^t \sigma g^{-1}.$$

In particular, the one-parameter subgroup V_1 acts on E in a unipotent way. Indeed, $v_1(t)\sigma = \exp(-t\tau)\sigma$, where

$$\tau\sigma = \frac{d}{dt}\Big|_{t=0} v_1(-t)\sigma = (v_1^t \sigma + \sigma v_1)$$

and $v_1 := \begin{pmatrix} 0 & 1 & 0 \\ 0 & 0 & 1 \\ 0 & 0 & 0 \end{pmatrix}$. One can verify that

$$\tau\left(\begin{pmatrix} a_1 & a_2 & a_3 \\ a_2 & a_4 & a_5 \\ a_3 & a_5 & a_6 \end{pmatrix}\right) = \begin{pmatrix} 0 & a_1 & a_2 \\ a_1 & 2a_2 & a_3 + a_4 \\ a_2 & a_3 + a_4 & 2a_5 \end{pmatrix}$$

and $\tau^5 = 0$. Let L be the set of the common fixed points of the $v_1(t)$'s. Thus,

$$L = \ker \tau = \mathbb{R}\sigma_0 + \mathbb{R}\sigma_1,$$

where $\sigma_0 := \begin{pmatrix} 0 & 0 & 1 \\ 0 & -1 & 0 \\ 1 & 0 & 0 \end{pmatrix}$ and $\sigma_1 := \begin{pmatrix} 0 & 0 & 0 \\ 0 & 0 & 0 \\ 0 & 0 & 1 \end{pmatrix}$. Observe that these matrices correspond to the quadratic forms Q_0 and Q_1, introduced previously. (Recall that we have $Q_1(x_1, x_2, x_3) = x_3^2$.) Observe also that, since $Q_0(v_2(t)p) = Q_0(p) + 2tQ_1(p)$,

$$V_2\sigma_0 = \sigma_0 + \mathbb{R}\sigma_1. \tag{1}$$

Moreover, note that

$$\sigma_0 + \mathbb{R}\sigma_1 = L \cap G. \tag{2}$$

We want to apply Lemma 4.1. In order to do so, take a sequence $\{r_i\}_i \subseteq R$ with $\lim_i r_i = e$ (such sequences exist, by the previous lemma), and define $M := \{r_i^{-1}\sigma_0\}_{i \in \mathbb{N}}$. Then $r_i^{-1}\sigma_0 \in L$ if and only if

$$r_i^{-1}\sigma_0 \in L \cap G = V_2\sigma_0,$$

that is, $r_i V_2 \in H$, or $r_i \in HV_2$. Hence, we may apply Lemma 4.1 if

$$\text{there exists } \{r_i\}_i \in R \setminus HV_2 \quad \text{such that} \quad \lim_{i \to \infty} r_i = e. \tag{3}$$

We shall have to distinguish two cases, according to whether (3) holds or not.

- *First case:* there exists a sequence $\{r_i\}_i \in R \setminus HV_2$ such that $\lim_{i \to \infty} r_i = e$. Then, by Lemma 4.1, there exist a non-constant polynomial $\psi : \mathbb{R} \to L$ with $\psi(0) = \sigma_0$ and a sequence $\{t_i\}_i \subseteq \mathbb{R}$ with

$$\lim_{i \to \infty} v_1(st_i)(r_i^{-1}\sigma_0) = \psi(s) \quad \forall s \in \mathbb{R}.$$

Note that $\psi(s) \in L \cap G$, that is, $\psi(s) \in \sigma_0 + \mathbb{R}\sigma_1$, by (2). Hence, there exists a non-constant polynomial $\varphi : \mathbb{R} \to \mathbb{R}$ with $\varphi(0) = 0$ and

$$\psi(s) = \sigma_0 + 2\varphi(s)\sigma_1.$$

Clearly, $\varphi(\mathbb{R})$ contains all positive or all negative real numbers. Thus, by (1),

$$\overline{V_1 R^{-1} H \sigma_0} \supset V_2^+ \sigma_0 \quad \text{or} \quad \overline{V_1 R^{-1} H \sigma_0} \supset V_2^- \sigma_0.$$

On the other hand, the orbit of σ_0 under G is the set of all symmetric indefinite matrices with determinant 1, Hence, it is closed in E and we have a homeomorphism

$$\theta : \mathrm{SL}(3, \mathbb{R})/H \to \mathrm{SL}(3, \mathbb{R}) \cdot \sigma_0, \quad gH \mapsto (g^{-1})^t \sigma_0 g^{-1}.$$

Therefore, $\overline{V_1 R^{-1} H}$ contains V_2^+ or V_2^- and the same is true for its inverse $\overline{HRV_1}$. Now apply Chap. IV, Lemma 2.2, to $X = \overline{Hx}, Y = Y, M = R, A = H$ and $B = C = V_1$ (recall that Y is compact). Then

$$\overline{Hx} \supset gY, \quad \forall g \in N_{\mathrm{SL}(3,\mathbb{R})}(V_1) \cap \overline{HRV_1}.$$

Observe that $N_{\mathrm{SL}(3,\mathbb{R})}(V_1)$ contains V_2^+ and V_2^-. Hence, $\overline{Hx}$ contains $V_2^+ y$ or $V_2^- y$, and the proof is finished in this case.

- *Second case:* there exists a neighbourhood U of e, such that

$$\{g \in G | \, gy \in \overline{Hx}\} \cap U \subseteq HV_2. \tag{4}$$

We are going to show that $\overline{Hx}$ contains V_2y. We first claim that there exists a sequence

$$\{r_i\}_i \subseteq R \cap V_2$$

with $\lim_i r_i = e$. Indeed, by the above lemma and our assumption, there is a sequence $\{g_i\}_i \subseteq R$ with $\lim_i g_i = e$ and $g_i = h_i r_i$ for $h_i \in H$ and $r_i = v_2(s_i) \in V_2, s_i \neq 0$. Then, using formula (5) from §2, for every $p \in \mathbb{R}^3$,

$$Q_0(g_i p) = Q_0(h_i r_i p) = Q_0(r_i p) = Q_0(p) + s_i Q_1(p).$$

As $g_i \to e$, this implies that $s_i \to 0$, that is, $r_i \to e$. As R is H-invariant, $r_i \in R$ and the claim is proved.

We know from Theorem 3.3 that the orbit of y under DV_1 is not closed. Hence, as X_1 is minimal under DV_1, by Chap. IV, Lemma 2.3, there exists a sequence $\{m_j\}_j \in G \setminus DV_1$ with

$$\lim_{j \to \infty} m_j = e \quad \text{and} \quad m_j y \in DV_1 y, \quad \forall j \in \mathbb{N}.$$

We may, of course, assume that $m_j \in U$.

Now, choose sequences $\{g_j\}_j$ in DV_1, $\{h_j\}_j$ in H and $\{v_j := v_2(t_j)\}_j$ in V_2 such that

$$m_j = h_j v_j \quad \text{and} \quad m_j y = g_j y, \quad \forall j \in \mathbb{N}. \tag{5}$$

(Observe that, by (4), $m_j \in HV_2$.) We first claim that the sequence $\{h_j\}_j$ is contained in DV_1.

Indeed, fix $j \in \mathbb{N}$ and observe that

$$g_j r_i g_j^{-1} m_j y = g_j r_i y \in g_j \overline{Hx} \in \overline{Hx}.$$

Hence, by (4), $g_j r_i g_j^{-1} m_j \in HV_2$, since $\lim_i g_j r_i g_j^{-1} m_j = m_j$ and so $g_j r_i g_j^{-1} m_j \in U$ for large i. Therefore, for large i,

$$r_i g_j^{-1} h_j = g_j^{-1}(g_j r_i g_j^{-1} m_j) v_j^{-1} \in H(HV_2)V_2 = HV_2.$$

As $r_i \in V_2 \setminus \{e\}$, Lemma 2.9 implies that $g_j^{-1} h_j$ and, hence, h_j is upper triangular.

On the other hand, since h_j is close to the identity, we may assume that its diagonal entries are positive. Hence, by Lemma 2.9 again, $h_j \in DV_1$. This proves the claim.

We claim that the discrete subgroup

$$\Delta := \{g \in DV | \, gy = y\}$$

of DV is contained in V. Indeed, since $h_j \in DV_1$, (5) shows that

$$g_j^{-1} h_j v_j \in \Delta.$$

On the other hand, the v_j's are distinct from e, since $m_j = h_j v_j \notin DV_1$. Moreover, $\lim_j v_j = e$. Hence, by Lemma 2.8 (ii), $\Delta \subseteq V$.

Therefore, $g_j^{-1} h_j v_2(t_j) \in V$ and, hence,

$$g_j^{-1} h_j \in DV_1 \cap V = V_1.$$

Thus,

$$v_2(t_j) y = h_j^{-1} g_j y \in V_1 Y = Y.$$

As V_2 centralizes V_1 and Y is V_1-minimal,

$$v_2(t_j) Y = v_2(t_j) \overline{V_1 y} = \overline{V_1 v_2(t_j) y} = Y.$$

But the set $\{t_j\}_{j \geq 0}$ is not discrete in $\mathbb{R}$ (since $v_j \to e$ and $v_j \neq e$). Hence, it generates a dense subgroup of $\mathbb{R}$. This implies that

$$V_2 Y = Y \subseteq \overline{Hx}.$$

Hence, $V_2 y \subseteq \overline{Hx}$. This completes the proof of the theorem. $\qquad\square$

5.2 Remark. If one is only interested in the weak form of Oppenheim's conjecture, then one can avoid the use of Margulis' Lemma in the above proof. Indeed, let Q be an indefinite irrational quadratic form in $n \geq 3$ variables. Assume, by contradiction, that there exists $\varepsilon > 0$ such that $Q(p) > \varepsilon$ for all $p \in \mathbb{Z}^n, p \neq 0$. Then, by Mahler's compactness criterion (see Chap. V, Theorem 3.2), $SO(Q)\mathbb{Z}^n$ is a relatively compact family of lattices in $\mathbb{R}^n$. Equivalently, the H–orbit of $g_Q \Gamma$ is relatively compact in G/Γ. This fact can be used instead of the arguments involving Margulis' Lemma in the above proof. Using the fact that $Hg_Q\Gamma$ is not closed, the conclusion is that $Hg_Q\Gamma$ contains $V_2^+ y$ or $V_2^- y$. Hence, $Q(\mathbb{Z}^n)$ is dense, and this is a contradiction.

5.3 Remark (A quantitative version of Oppenheim's conjecture). Let Q be an indefinite irrational quadratic form in $n \geq 3$ variables. We now know that, for any $a < b$, there exists $p \in \mathbb{Z}^n$ with $a < Q(p) < b$. It is natural to ask about the distribution of the values of Q on $\mathbb{Z}^n$. The following counting problem may be viewed as a quantitative version of Oppenheim's conjecture. For $T > 0$, let

$$V_{a,b}(\mathbb{Z}) = \{p \in \mathbb{Z}^n \,|\, a < Q(p) < b\},$$

and let

$$N_{a,b}(T) = |V_{a,b}(\mathbb{Z}) \cap B_T(0)|,$$

the number of points p in $V_{a,b}(\mathbb{Z})$ with $\|p\| < T$. What is the asymptotic behaviour of $N_{a,b}(T)$ when $T \to \infty$?

Let (p, q) be the signature of Q, $p \geq q$. The following remarkable results are proved in [EMM], Theorem 2.1, Theorem 2.2 and Theorem 2.3.

If $(p, q) \neq (2, 1)$ or $(2, 2)$, then

$$N_{a,b}(T) \sim \lambda(b-a)T^{n-2} \quad \text{as} \quad T \to \infty, \qquad (*)$$

where the constant λ is such that

$$V_{a,b}(T) := \text{vol}(\{p \in \mathbb{R}^n \,|\, a < Q(p) < b, \, \|p\| < T\} \sim \lambda(b-a)T^{n-2},$$

as $T \to \infty$. (The asymptotically exact lower bound was already proved in [DM4].) Surprisingly, the above asymptotic formula $(*)$ does not hold longer if $(p, q) = (2, 1)$ or $(2, 2)$. In fact, for every $\varepsilon > 0$ and every $a < b$, there exists an irrational form Q of signature $(2, 1)$ or $(2, 2)$ and a constant $c > 0$ such that for a sequence T_i with $T_i \to \infty$

$$\frac{N_{a,b}(T_i)}{V_{a,b}(T_i)} > c(\log T_i)^{1-\varepsilon}.$$

However, for any Q of signature $(2, 1)$ or $(2, 2)$ and any $a < b$, there exists an upper bound for $N_{a,b}(T)$ of the form $cT^{n-2}\log T$.

§6 Ratner's Results on the Conjectures of Raghunathan, Dani and Margulis

Let G be a connected Lie group, and let Γ be a lattice in G. Look at the action of a closed subgroup H of G on G/Γ. In general, the orbits of H are "chaotic" (if H acts ergodically, then almost all of them are dense; see Chap. IV, Proposition 1.5). However, we saw some examples where the closures of all orbits were homogeneous spaces themselves. This is, for instance, the case in Kronecker's theorem about translations on a torus (see Chap. I, Example 1.12 (iv)), or in Hedlund's Minimality Theorem (Chap. IV, Theorem 1.9). Another example is the Dani–Margulis result mentioned at the beginning of this chapter. One may wonder how general this phenomenon is. In the 1980's, some conjectures were formulated in this respect.

A closed subgroup U of G is called Ad-*unipotent* if Ad u, acting on the Lie algebra of G, is unipotent for all $u \in U$.

Conjecture 1 (Raghunathan). *Let G be a connected Lie group, and let Γ be a lattice in G. Let U be an Ad-unipotent subgroup of G. Then, for any $x \in G/\Gamma$, the closure of the U-orbit Ux is homogeneous, that is, there exists a closed subgroup P of G containing U such that $\overline{Ux} = Px$.*

This conjecture, first formulated in [Da4], is due to Raghunathan who noted its connection with Oppenheim's conjecture. For $G = \text{SL}(2, \mathbb{R})$, all unipotent connected subgroups are conjugated to the subgroup N of the upper triangular unipotent matrices. So, by Hedlund's Minimality Theorem, Conjecture 1 is true in this case.

Conjecture 1 was proved for so–called horospherical subgroups U by Dani in the case of a reductive Lie group G [Da6]. (A subgroup U is called horospherical if there exists $a \in G$ such that

$$U = \{g \in G \mid a^i g a^{-i} \to e \quad \text{as} \quad i \to \infty\}.$$

Starkov [St1] solved the same conjecture in the case of a solvable Lie group. The first results on non-horospherical subgroups of semisimple Lie groups were obtained in [DM2] for $\mathrm{SL}(3, \mathbb{R})$.

In his article [Ma2] solving Oppenheim's conjecture, Margulis stated the following strengthening of Raghunathan's conjecture.

Conjecture 2 (Margulis). *Let G be a connected Lie group, and let Γ be a lattice in G. Let H be a subgroup which is generated by* Ad *-unipotent elements. Then, for any $x \in G/\Gamma$, the closure of the H-orbit Hx is homogeneous.*

Observe that the class of subgroups covered by Conjecture 2 is much larger than the class of Ad-unipotent subgroups. It contains, for example, all non-compact simple Lie groups. As mentioned in the introduction, Conjecture 2 was solved in [DM1] for $G = \mathrm{SL}(3, \mathbb{R})$ and $H = \mathrm{SO}(2, 1)$.

The question of distribution of orbits involves the study of invariant measures. A natural class of probability measures on G/Γ arises as follows. Let H be a closed subgroup such that $H \cap g \Gamma g^{-1}$ is a lattice in H, for some $g \in G$. The H-invariant measure ν_H on

$$H/H \cap g \Gamma g^{-1} \cong Hx$$

then induces an H-invariant measure on G/Γ supported on the orbit Hx of $x = g\Gamma$. Observe that this forces the orbit Hx to be closed (see [Rag], Theorem 1.13). Measures on G/Γ of this type are called *algebraic measures*, that is, a probability measure μ on G/Γ is algebraic if there is a closed subgroup H of G such that μ is H–invariant and supp $\mu = Hx$ for some $x \in G/\Gamma$.

In the above mentioned paper [Da6], Dani formulated (in a slightly weaker form) the following conjecture about the invariant measures under Ad-unipotent subgroups.

Conjecture 3 (Dani). *Let G be a connected Lie group, and let Γ be a lattice in G. Let U be an* Ad *-unipotent subgroup of G. Then every ergodic and U-invariant Borel probability measure on G/Γ is algebraic.*

In the case $G = \mathrm{SL}(2, \mathbb{R})$, this conjecture follows from the unique ergodicity result of Furstenberg [Fu1] for a uniform lattice Γ and from Dani's classification result [Da1] (see Chap. IV, §3).

Let $\{\varphi_t\}_t$ be a one-parameter subgroup of homeomorphisms of a locally compact X. The orbit of $x \in X$ under $\{\varphi_t\}_t$ is *uniformly distributed* with respect to a probability measure μ on X, if for every bounded continuous function f on X,

$$\frac{1}{T} \int_0^T f(\varphi_t x) dt \to \int_X f d\mu \quad \text{as} \quad T \to \infty.$$

As discussed in Chapter IV, §4, Dani and Smillie [DS] proved that, for $G = \mathrm{SL}(2, \mathbb{R})$ and $\{u_t\}$ a unipotent flow, every non-periodic $\{u_t\}$-orbit in G/Γ is uniformly distributed with respect to the G –invariant measure. The following conjecture was formulated by Dani [Da5] for $G = \mathrm{SL}(n, \mathbb{R}), \Gamma = \mathrm{SL}(n, \mathbb{Z})$ and by Margulis [Ma1] in general.

Conjecture 4 (Dani, Margulis). *Let G be a connected Lie group, and let Γ be a lattice in G. Let $\{u_t\}$ be an Ad -unipotent one-parameter subgroup of G. Then, for every point $x \in G/\Gamma$, the orbit $\{u_t x\}$ is uniformly distributed with respect to a probability measure μ_x on G/Γ.*

In what constitutes a *tour de force*, M. Ratner solved, in the early 1990's, all these conjectures in full generality. In fact, she proved even more general results. Let us first state her measure classification theorem ([Ra2], [Ra3], [Ra4]).

6.1 Theorem (Classification of Invariant Measures). *Let G be a connected Lie group, and let Γ be a discrete subgroup of G (not necessarily a lattice). Let H be a connected Lie subgroup of G generated by Ad-unipotent one-parameter groups. Then every ergodic H-invariant Borel probability measure on G/Γ is algebraic.*

Using this result and the quantitative version of Margulis' Lemma (see Chap. IV, Theorem 4.2, Chap. V, Theorem 5.1, as well as the Notes of Chapter V), Ratner deduced the following equidistribution theorem in [Ra5].

6.2 Theorem (Uniform Distribution). *Let G be a connected Lie group, and let Γ be a lattice in G. Let $\{u_t\}$ be an Ad -unipotent one-parameter subgroup of G. Then, for any $x \in G/\Gamma$, the orbit $\{u_t x\}$ is uniformly distributed with respect to an algebraic measure μ_x on G/Γ.*

The corresponding version of this theorem for unipotent cyclic subgroups is also proved in [Ra5]. Following [Ra7], we described in Chap. IV, §3 and §4 some ingredients of Ratner's original proof of the above theorems in the case $G = \mathrm{SL}(2, \mathbb{R})$. An overview of the general proof is given in [Ra10]. For a nilpotent Lie group G, the above equidistribution result was proved by Lesigne [Les1], [Les2].

Observe that Ratner's uniform distribution theorem immediately implies Raghunatan's conjecture (Conjecture 1) for one-parameter unipotent subgroups. In fact, Ratner [Ra5] used the uniform distribution theorem to prove the following strengthening of Margulis' conjecture.

6.3 Theorem (Orbit Closures). *Let G be a connected Lie group, and let Γ be a lattice in G. Let H be a connected Lie subgroup of G generated by Ad -unipotent one-parameter groups. Then, for any $x \in G/\Gamma$, there exists a closed connected subgroup P containing H such that $\overline{Hx} = Px$ and Px admits a P –invariant probability measure.*

Let us indicate how Oppenheim's conjecture follows from the result about orbit closures. Let Q be an indefinite irrational quadratic form in $n \geq 3$ variables, and let $H = \mathrm{SO}(Q)$. Since $n \geq 3$, the connected component H_0 of H is generated by unipotent elements. Therefore, by the above theorem, the closure of the orbit $H_0\Gamma$ of $\Gamma = \mathrm{SL}(n, \mathbb{Z})$ in G/Γ is equal to $P\Gamma$ for a closed connected subgroup P of $G = \mathrm{SL}(n, \mathbb{R})$ containing H_0. On the other hand, H_0 is a maximal connected subgroup of G. Hence, $P = H_0$ or $P = G$. In the first case, $H\Gamma$ is closed (since H_0 has finite index in H) and, hence, Q is rational (see Proposition 2.4). In the second case, $H\Gamma$ is dense and, hence, $Q(\mathbb{Z}^n)$ is dense in $\mathbb{R}$.

6.4 Remark. In the above mentioned papers, Ratner proved that the classification of invariant measures and the result about orbit closures (Theorems 6.1 and 6.3) hold – more generally – if the subgroup H is of the form $H = \cup_{i=1}^{\infty} u_i H_0$, where all u_i are Ad -unipotent elements in G, H/H_0 is finitely generated, and the connected component H_0 of the identity in H is generated by Ad -unipotent one-parameter groups. Extending these results, N. Shah showed in a recent work [Sh2] that Theorems 6.1 and 6.3 are valid for an arbitrary subgroup H generated by Ad -unipotent elements.

Notes

A. Oppenheim [Op1] formulated his conjecture (in its weak form) for $n \geq 5$. This restriction was suggested by Meyer's theorem (see [Se]) according to which any rational indefinite form in $n \geq 5$ variables represents zero non-trivially over $\mathbb{Z}$, a fact which is no longer true if $n < 5$. Oppenheim's conjecture for $n \geq 3$ was stated for diagonal forms in [DaH] and proved for $n \geq 5$. A. Davenport, together with B. Birch [BD] and H. Ridout [DaR], succeeded in proving the conjecture for large n. Other partial results were obtained by Watson [Wa1], [Wa2], Baker and Schlickewey [BS] and Iwaniec [Iw]. The methods used so far were methods from analytic number theory. The idea of attacking Oppenheim's conjecture via flows on homogeneous space is due to Raghunathan who noticed the relationship with dynamical systems. In implicit form, this relationship appeared already in a 1955 article by Cassels and Swinnerton-Dyer [CS]. For a precise and comprehensive survey of results and methods connected with Oppenheim's conjecture, with historical details, see the article [Ma7] by Margulis.

The analogue of Oppenheim's conjecture in the p-adic and S-arithmetic cases was proved by Borel and Prasad (see [BoP1], [BoP2]).

An excellent survey on Ratner's work is Ghys' report in Séminaire Bourbaki [Gh]. Two other nice and more recent surveys covering a wide range of topics around flows on homogeneous spaces are [Da8] and [St3].

Mozes and Shah [MS] showed, using Ratner's invariant measure theorem, that the set of probability measures on G/Γ which are invariant under some (not fixed) unipotent one-parameter subgroup is a closed subset of the set of all measures. Another application of this theorem is the rigidity theorem of Ratner and Witte for unipotent translations (see [Ra1], [Wi], and also [Ra10] and [Gh]).

Extending Ratner's uniform distribution theorem (Theorem 6.2), Shah [Sh1] showed that, for a polynomial curve $\{\theta(t)\}_t$ in a closed subgroup G of $\mathrm{SL}(n, \mathbb{R})$ and a lattice Γ in G, the trajectory $\{\theta(t)\Gamma\}_t$ is uniformly distributed with respect to an algebraic measure on G/Γ. This nice result is much in the spirit of Weyl's classical theorem on uniform distribution of polynomials (see [CFS], Chap. 7, §2).

Eskin, Mozes and Shah give in [EMS2] a non-divergence theorem for translates of algebraic measures on G/Γ when G is a reductive algebraic group and Γ an arithmetic lattice in G. The first step in the proof is a generalization of the results of Dani and Margulis on the recurrence properties of unipotent trajectories to a compact subset of G/Γ (see Chapter V, §5).

Ratner ([Ra8], [Ra9]) has extended her results to the p-adic and the S-arithmetic cases. A shorter proof of the invariant measures theorem has been given by Margulis and Tomanov [MT] in the case of an algebraic group G.

Dani established in [Da11] a correspondence between Diophantine approximation of a family of m vectors in $\mathbb{R}^n$ and behaviour at infinity of certain orbits in the space $\mathrm{SL}(n + m, \mathbb{R})/\mathrm{SL}(n+m, \mathbb{Z})$ of the unimodular lattices in $\mathbb{R}^{n+m}$. An exciting recent development is the new approach based on Dani's correspondence of Kleinbock and Margulis ([KM2]) to the theory of Diophantine approximation on manifolds. Among other things, this approach allows them to solve conjectures formulated in the 1970's by Baker and Sprindžuk.

Bibliography

[AA]   V.I. Arnold, A. Avez: *Ergodic Problems of Classical Me-chanics*, Addison-Wesley, 1988.

[AG]   L. Auslander, L. Green: *G*–induced flows, *Amer. J. Math.* **88**, (1966), 43–60.

[AGH]   L. Auslander, J. Green, F. Hahn: *Flows on Homogeneous Spaces*, Annals of Math. Studies 53, Princeton Univ. Press, 1963.

[An]   D.V. Anosov: *Geodesic Flows on Riemannian Manifolds of Negative Curvature*, Proc. Sketlov Inst. Math. 90, Amer. Math. Soc., 1967.

[Ba]   W. Ballmann: *Lectures on Spaces of Nonpositive Curvature*, Birkhäuser, 1995.

[Bar]   H.-J. Bartels: Nichteuklidische Gitterpunktprobleme und Gleichverteilung in linearen algebraischen Gruppen, *Comment. Math. Helv.* **57**, (1982), 158–172.

[BD]   B.J. Birch, H. Davenport: Quadratic equations in several variables, *Proc. Cambridge Phil. Soc.* **54**, (1958), 135–138.

[Be]   A. Beardon: *The Geometry of Discrete Groups*, Springer Verlag, 1983.

[BeM]   M.B. Bekka, M. Mayer: On Kazhdan's property (T) and Kazhdan's constants associated to a Laplacian on $SL(3, \mathbb{R})$, *J. Lie Theory* **10**, (2000), 93–105.

[Bir]   G. D. Birkhoff: Proof of the ergodic theorem, *Proc. Acad. Sci. USA* **17**, (1931), 656–660.

[BKS]   T. Bedford, M. Keane, C. Series: *Ergodic Theory, Symbolic Dynamics and Hyperbolic Spaces*, Oxford University Press, 1992.

[BL]   M. Babillot, F. Ledrappier: Geodesic paths and horocycle flow on abelian covers, in: *Proc. International Colloquium on Lie Groups and Ergodic Theory*, Tata Institute of Fundamental Research, Narosa Publishing House, New Delhi 1998, 1–32.

[BM]   J. Brezin, C.C. Moore: Flows on homogeneous spaces: A new look, *Amer. J. Math.* **103**, (1981), 571–613.

[Bo1]   N. Bourbaki: *Intégration, Chap. 6*, Hermann, 1963.

[Bo2]   N. Bourbaki: *Intégration Chap. 7–8*, Hermann, 1963.

[BoP1]   A. Borel, G. Prasad: Valeurs de formes quadratiques aux points entiers, *C. R. Acad. Sci. Paris, Série I*, **307**, (1988), 217–220.

[BoP2]   A. Borel, G. Prasad: Values of isotropic quadratic forms at *S*-integral points, *Compositio Math.* **83**, (1992), 347–372.

[Bor]   A. Borel: *Introduction aux Groupes Arithmétiques*, Hermann, 1969.

[Bow]   R. Bowen: Weak mixing and unique ergodicity on homogeneous spaces, *Isr. J. Math.* **23**, (1976), 267–273.

[BP]   M. Babillot, M. Peigné: Closed geodesic in homology classes of hyperbolic manifolds with cusps, *C. Acad. Sci. Paris, Série I* **324**, (1997), 901–906.

[BrD]   T. Bröcker, T. tom Dieck: *Representations of Compact Lie Groups* , Springer Verlag, 1985.

[BS]   R.C. Baker, H.P. Schlickewey: Indefinite quadratic forms, *Proc. London Math.Soc.* **54**, (1987), 383–411.

[Bu]   M. Burger: Horocycle flows on geometrically finite surfaces, *Duke Math. J.* **61**, (1990), 779–803.

[BuS]   M. Burger, V. Schroeder: Volume, diameter and the first eigenvalue of locally symmetric spaces of rank one, *J. Diff. Geom.* **26**, (1987), 273–289.

[BW]   A. Borel, N. Wallach: *Continuous Cohomology, Discrete Groups and Representations of Reductive Groups*, Annals of Math. Studies 94, Princeton Univ. Press, 1980.

[CaM]   W. Casselman, D. Milicic: Asymptotic behaviour of matrix coefficients of admissible representations, *Duke Math. J.* **49**, (1982), 73–96.

[CFS]   I. P. Cornfeld, S. V. Fomin, Ya. G. Sinai: *Ergodic Theory*, Springer Verlag, 1982.

[CHH]   M. Cowling, U. Haagerup, R. Howe: Almost L^2 matrix coefficients, *J. Reine Angew. Math.* **387**, (1988), 97–110.

[Co]   M. Cowling: Sur les coefficients des représentations unitaires des groupes de Lie simples, *Springer Lecture Notes in Math.* **739**,(1979), 132–178.

[CS]   J.W.S. Cassels, H.P.F. Swinnerton-Dyer: On the product of three homogeneous forms and indefinite ternary quadratic forms, *Phil. Trans. Roy. Soc. London* **248**, (1955), 869–930.

[Da1]   S.G. Dani: Invariant measures of horospherical flows on non compact homogeneous spaces, *Invent. Math.* **47**, (1978) 101–138.

[Da2]   S.G. Dani: On orbits of unipotent flows on homogeneous spaces, *Ergod. Th. & Dynam. Syst.* **4**, (1984), 25–34.

[Da3]   S.G. Dani: On orbits of unipotent flows on homogeneous spaces II, *Ergod. Th. & Dynam. Syst.* **6**, (1986), 167–182.

[Da4]   S.G. Dani: Invariant measures and minimal sets of horospherical flows, *Invent. Math.* **64**, (1981), 357–385.

[Da5]   S.G. Dani: On uniformly distributed orbits of horocycle flows, *Ergod. Th. & Dynam. Syst.* **2**, (1982), 139–158.

[Da6]   S.G. Dani: Orbits of horospherical flows, *Duke Math. J.* **53**, (1986), 177–188.

[Da7]   S.G. Dani: On invariant measures, minimal sets and a lemma of Margulis, *Invent. Math.* **51**, (1979), 239–260.

[Da8]   S.G. Dani: Flows on homogeneous spaces: A review, in: Proc. of the Warwick Symposium Ergodic theory on $\mathbb{Z}^d$-actions, London Math. Soc., *Cambridge University Press, Lectures Notes Series* **228**,(1996), 63–122.

[Da9]   S.G. Dani: A proof of Margulis' theorem on values of quadratic forms, independent of the axiom of choice, *Enseig. Math.* **40**, (1994), 49–58.

[Da10]   S.G. Dani: Bernoullian translations and minimal horospheres on homogeneous spaces, *J. Indian Math. Soc.* **40**, (1976), 245–284.

[Da11]   S.G. Dani: Divergent trajectories of flows on homogeneous spaces and Diophantine approximation, *J. Reine Angew. Math.* **359**, (1985), 55–87.

[DaH]   H. Davenport, H. Heilbronn: On indefinite quadratic forms *Proc. London Math. Soc.* **21**, (1946), 185–193.

[DaR]   H. Davenport, D. Ridout: Indefinite quadratic forms, *Proc. London Math. Soc. (3)* **9**, (1959), 544–555.

[De1]   J. Delsarte: Sur le gitter fuchsien, *C.R. Acad. Sci. Paris* **214**, (1942), 147–149.

[De2]   J. Delsarte: *Oeuvres,Tome II*, Édition du CNRS, Paris 1971.

[DM1]   S.G. Dani, G.A. Margulis: Values of quadratic forms at primitive integral points, *Invent. Math.* **98**, (1989), 405–424.

[DM2]   S.G. Dani, G.A. Margulis: Orbit closures of generic unipotent flows on homogeneous spaces of $SL(3, \mathbb{R})$, *Math. Ann.* **286**, (1990), 101–128.

[DM3]   S.G. Dani, G.A. Margulis: Values of quadratic forms at primitive integral points: An elementary approach, *Enseig. Math.* **36**, (1990), 143–174.

[DM4]   S.G. Dani, G.A. Margulis: Limit distributions of orbits of unipotent flows and values of quadratic forms, *Adv. in Soviet Math.* **16**, (1993), 91–137.

[DRS]   W. Duke, Z. Rudnik, P. Sarnak: Density of integer points on affine homogeneous varieties, *Duke Math. J.* **71**, (1993), 143–180.

[DS]   S.G. Dani, J. Smillie: Uniform distribution of horocycle orbits for Fuchsian groups, *Duke Math. J.* **51**, (1984), 185–194.

[EM]   A. Eskin, C. McMullen: Mixing, counting and equidistri-
            bution on Lie groups, *Duke Math. J.* **71**, (1993), 181–209.

[EMM]   A. Eskin, G.A. Margulis, S. Mozes: Upper bounds and
            asymptotics in a quantitative version of the Oppenheim
            conjecture, *Ann. Math.* **147**, (1998), 93–141.

[EMS1]   A. Eskin, S. Mozes, N. Shah: Unipotent flows and count-
            ing lattice points on homogeneous varieties, *Ann. Math.*
            143, (1996), 253–299.

[EMS2]   A. Eskin, S. Mozes, N. Shah: Non divergence of translates
            of certain algebraic measures, *Geom. Anal. and Funct.
            Anal.* **7**, (1997), 48–80.

[EP]   A. Ellis, W. Perrizo: Unique ergodicity of flows on homo-
            geneous spaces, *Isr. J. Math.* **29**, (1978), 276–284.

[FK]   H. M. Farkas, I. Kra: *Riemann Surfaces*, Springer Verlag,
            1992.

[Fu1]   H. Furstenberg: The unique ergodicity of the horocyclic
            flow, *Springer Lecture Notes in Math.* **318**,(1972), 95–115.

[Fu2]   H. Furstenberg: *Recurrence in Ergodic Theory and Combi-
            natorial Number Theory*, Princeton Univ. Press, 1981.

[GF]   I. Gelfand, S.V. Fomin: Geodesic flows on manifolds of
            constant negative curvature, *Transl. II. Ser., Am. Math.
            Soc.* **1**, (1955), 49–65.

[GGP]   I.M. Gelfand, M.I. Graev, I.I. Pyatetskii–Shapiro: *Repre-
            sentation Theory and Automorphic Functions*, Saunders,
            1969.

[Gh]   E. Ghys: Dynamique des flots unipotents sur les espaces
            homogènes, *Astérisque* **206**, (1992), 93–136.

[GHL]   S. Gallot, D. Hulin, J. Lafontaine: *Riemannian Geometry*,
            Springer Verlag, 1987.

[Gün]   P. Günther: Problèmes de réseaux dans les espaces hy-
            perboliques, *C. R. Acad. Sci. Paris, Sér A* **288**, (1979),
            49–52.

[Ha]   J. Hadamard: Les surfaces à courbures opposées et leurs
            lignes géodésiques, *J. Math. Pures Appl.* **4**, (1898), 27–74.

[Hal]   P. Halmos: *Lectures on Ergodic Theory*, Chelsea, 1953.

[He1]   G. Hedlund: Dynamics of geodesic flows, *Bull. Am. Math.
            Soc.* **45**, (1939), 241–260.

[He2]   G. Hedlund: Fuchsian Groups and Mixtures, *Ann. of
            Math.* **40**, (1939), 370–383.

[Hel]   S. Helgason: *Differential Geometry and Symmetric Spaces*,
            Academic Press, 1962.

[HeR]   E. Hewitt, K. Ross: *Abstract Harmonic Analysis I*, Springer
            Verlag, 1979.

[HM]   R. Howe, C.C. Moore: Asymptotic properties of unitary
            representations, *J. Funct. Anal.* **32**, (1979), 72–96.

[Ho1]   E. Hopf: Fuchsian groups and ergodic theory, *Trans. Am. Math. Soc.* **39**, (1936), 299–314.

[Ho2]   E. Hopf: Statistik der geodätischen Linien auf Mannig-faltigkeiten negativer Kruemmung, *Ber. Verh. Sachs. Akad. Wiss. Leipzig* **91**, (1939), 261–304.

[How]   R. Howe: On a notion of rank for unitary representations of the classical groups, in: A. Figà Talamanca, ed., *Harmonic Analysis and Group Representations*, CIME II, Liguori editore, (1982), 223–332.

[HT]   R. Howe, Eng Che Tang: *Non Abelian Harmonic Analysis*, Springer Verlag, 1992.

[Hub]   H. Huber: "Uber eine neue Klasse automorpher Funktionen und ein Gitterpunktproblem in der hyperbolischen Ebene, *Comment. Math. Helv.* **30**, (1956), 20–62.

[Iw]   H. Iwaniec: On indefinite quadratic forms in four variables, *Acta. Arith.* **33**, (1977), 209–229.

[Ka]   S. Katok: *Fuchsian Groups*, Chicago Lecture Notes in Math. Univ. of Chicago Press, 1992.

[KH]   A. Katok, B. Hasselblatt: *Introduction to the Modern Theory of Dynamical Systems*, Cambridge Univ. Press, 1995.

[KM1]   D.Y. Kleinbock, G.A. Margulis: Bounded orbits of non-quasiunipotent flows on homogeneous spaces, *Amer. Math. Soc. Transl.* **171**, (1996), 141–172.

[KM2]   D.Y. Kleinbock, G.A. Margulis: Flows on homogeneous spaces and Diophantine approximation on manifolds, *Ann. Math.* **148**, (1998), 339–360.

[La]   S. Lang: *Algebra*, Addison-Wesley, 1965.

[Ld]   F. Ledrappier: Un champ Markovien peut être d'entropie nulle et mélangeant, *C. R. Acad. Sci. Paris, Série A* **2807**, (1978), 561–562.

[Le]   J. Lehner: *Discontinuous Groups and Automorphic Functions*, AMS Providence, 1964.

[Les1]   E. Lesigne: Théorèmes ergodiques pour une translation sur une nilvarieté, *Ergod. Th. & Dynam. Syst.* **9**, (1989), 115-126.

[Les2]   E. Lesigne: Sur une nilvarieté, les parties minimales associées à une translation sont uniquement ergodiques, *Ergod. Th. & Dynam. Syst.* **11**, (1991), 379–391.

[LP]   P. Lax, R. Phillips: The asymptotic distribution of lattice points in euclidean an non-euclidean spaces, *J. Funct. Anal.* **46**, (1982), 280–350.

[Ma1]   G.A. Margulis: Lie groups and ergodic theory, *Springer Lecture Notes in Math.* **1352**, (1989), 130–146.

[Ma2]   G.A. Margulis: Formes quadratiques indéfinies et flots unipotents sur les espaces homogènes, *C. R. Acad. Sci.*

Paris, Série I **304**, (1987), 249–253.

[Ma3]   G.A. Margulis: Discrete subgroups and ergodic theory, in: *Number theory, Trace Formulas, Discrete Groups*, Symposium in honor of A. Selberg, Acad. Press 1989, 377–398.

[Ma4]   G.A. Margulis: On some applications of ergodic theory to the study of manifolds of negative curvature, *Funct. Anal. Appl.* **3**, (1969), 335–336.

[Ma5]   G.A. Margulis: On the action of unipotent groups in the space of lattices, in: *Proc. of the Summer School on Group Representations*, Bolyai Janos Math. Soc., Budapest, 1971, 365–371.

[Ma6]   G.A. Margulis: Non uniform lattices in semisimple algebraic groups, in: *Proc. of the Summer School on Group Representations*, Bolyai Janos Math. Soc., Budapest, 1971, 371–553.

[Ma7]   G.A. Margulis: Oppenheim conjecture, in: *Fields Medalists' Lectures*, World Scientific, Singapore Univ. Press 1997, 272–327.

[Mah1]   K. Mahler: On lattice points in n–dimensional star bodies: I. Existence theorems, *Proc. Roy. Soc. London* **A 187**, (1946), 151–187.

[Mah2]   K. Mahler: An arithmetic property of groups of linear transformations, *Acta Arith.* **5**, (1959), 197–203.

[Man]   R. Mañé: *Ergodic Theory and Differential Dynamics*, Springer Verlag, 1987.

[Mau]   F.I. Mautner: Geodesic flows on symmetric Riemannian spaces, *Ann. Math.* **65**, (1957), 416–430.

[Mc]   G. W. Mackey: *Unitary Group Representations in Physics, Probability and Number Theory*, Benjamin, 1978.

[Mi]   H. Minkowski: Diskontinuitätsbereich für arithmetische Äquivalenz, *J. Reine Ang. Math.* **129**, (1905), 220–274.

[Mo1]   C.C. Moore: Ergodicity of homogeneous flows, *Amer. J. Math.* **88**, (1966), 154–178.

[Mo2]   C.C. Moore: The Mautner phenomenon for general unitary representations, *Pacific J. Math.* **86**, (1980), 155–169.

[Mo3]   C.C. Moore: Exponential decay of correlation coefficients for geodesic flows, in: *Group representations, Ergodic Theory, Operator Algebras and Mathematical Physics*, Proc. Conference in honor G. W. Mackey, MSRI Publications, Springer Verlag 1987, 163–181.

[Mor]   M. Morse: Recurrent geodesics on a surface of negative curvature, *Trans. Amer. Math. Soc.* **22**, (1921), 84–100.

[Mos]   G. D. Mostow: *Strong Rigidity of Locally Symmetric Spaces*, Annals of Math. Studies, Princeton Univ. Press, 1973.

[Mr]   B. Marcus: The horocycle flow is mixing of all orders,
              Invent. Math. **46**, (1978), 201–209.

[MS]   S. Mozes, N. Shah: On the space of ergodic invariant
              measures of unipotent flows, *Ergod. Th. & Dynam. Syst.*
              15, (1995), 149–159.

[MT]   G.A. Margulis, G.M. Tomanov: Invariant measures for ac-
              tions of unipotent groups over local fields on homogeneous
              spaces, *Invent. Math.* **116**, (1994), 347–392.

[Mz1]   S. Mozes: Mixing of all orders of Lie group actions, *Invent.*
              Math. **107**, (1992), 235–241.

[Mz2]   S. Mozes: Erratum: Mixing of all orders of Lie group
              actions, *Invent. Math.* **119**, (1995), 399.

[Op1]   A. Oppenheim: The minima of the indefinite quaternary
              quadratic forms, *Ann. Math.* **32**, (1931), 271–298.

[Op2]   A. Oppenheim: Values of quadratic forms I, *Quart. J.*
              Math., Oxford Ser. (2) **4**, (1953), 54–59.

[Op3]   A. Oppenheim: Values of quadratic forms II, *Quart. J.*
              Math., Oxford Ser. (2) **4**, (1953), 60–66.

[Op4]   A. Oppenheim: Values of quadratic forms III, *Monatshefte*
              Math. Phys **57**, (1953), 97–101.

[OW]   D. Ornstein, B. Weiss: Geodesic flows are Bernoullian,
              Isr. J. Math. **14**, (1973), 184–198.

[Pa]   G. Pansu: Le flot géodesique des variétes Riemanniennes
              à courbure négative, *Séminaire Bourbaki* **738**, (1991), 269–
              298.

[Pat]   S. J. Patterson: A lattice-point problem in hyperbolic
              space, *Mathematika* **22**, (1975), 81–88.

[PhR]   R. Phillips, Z. Rudnick: The circle problem in the hyper-
              bolic plane, *J. Funct. Anal.* **121**, (1994), 78–116.

[Pi]   J.-P. Pier: *Amenable locally compact groups*, Wiley, 1984.

[Pr1]   W. Parry: Ergodic properties of affine transformations
              and flows on nilmanifolds, *Amer. J. Math.* **91**, (1969),
              757–771.

[Pr2]   W. Parry: Metric classification of ergodic nilflows, *Amer.*
              J. Math. **93**, (1971), 819–828.

[PS]   M. Pollicott, R. Sharp: Orbit counting for some discrete
              groups acting on simply connected manifolds with negative
              curvature, *Invent. Math.* **117**, (1994), 275–302.

[Pt]   A. Paterson: *Amenability*, AMS Providence, 1988.

[Ra1]   M. Ratner: Ergodic theory in hyperbolic space, *Contemp.*
              Math **26**, (1984), 309–334.

[Ra2]   M. Ratner: Strict measure rigidity for unipotent sub-
              groups of solvable groups, *Invent. Math.* **101**, (1990),
              449–482.

[Ra3]   M. Ratner: On measure rigidity of unipotent subgroups of
             semi simple groups, *Acta Math.* **165**, (1990), 229–309.

[Ra4]   M. Ratner: On Raghunathan's conjecture, *Ann. Math.*
             134, (1991), 545–607.

[Ra5]   M. Ratner: Raghunathan's topological conjecture and dis-
             tributions of unipotent flows, *Duke J. Math.* **63**, (1991),
             235–280.

[Ra6]   M. Ratner: Distribution rigidity for unipotent actions on
             homogeneous spaces, *Bull. AMS* **24**, (1991), 321–325.

[Ra7]   M. Ratner: Raghunathan's conjectures for $SL(2,\mathbb{R})$, *Isr.
             J. Math.* **80**, (1992), 1–31.

[Ra8]   M. Ratner: Raghunathan's conjectures for p-adic Lie
             groups, *Intern. Math. Res. Notices* **5**, (1993), 141–146.

[Ra9]   M. Ratner: Raghunathan's Conjectures for cartesian prod-
             ucts of real and p-adic groups, *Duke J. Math.* **77**, (1995),
             275–382.

[Ra10]   M. Ratner: Invariant measures and orbit closures for
             unipotent actions on homogeneous spaces, *Geom. and Funct
             Anal.* **4**, (1994), 236–257.

[Ra11]   M. Ratner: The rate of mixing for geodesic and horocycle
             flows, *Ergod. Th. & Dynam. Syst.* **7**, (1987), 267–288.

[Rag]   M. S. Raghunathan: *Discrete Subgroups of Lie Groups*,
             Springer Verlag, 1972.

[Rd]   W. Rudin: *Real and Complex Analysis*, McGraw Hill, 1983.

[Re]   H. Reiter: *Classical Harmonic Analysis and Locally Com-
             pact Groups*, Oxford University Press, 1968.

[Ro]   T. Roblin: *Sur la théorie ergodique des groupes discrets en
             géométrie hyperbolique*, Thesis, Université d'Orsay, 1999.

[Sch1]   K. Schmidt: *Dynamical Systems of Algebraic Origin*,
             Birkhäuser, 1995.

[Sch2]   K. Schmidt: Asymptotic properties of unitary representa-
             tions and mixing, *Proc. London Math. Soc.* **48**, (1984),
             445–460.

[Se]   J.-P. Serre: *A Course in Arithmetic*, Springer Verlag, 1978.

[Sg1]   C.L. Siegel: *Lectures on the Geometry of Numbers*, Springer
             Verlag, 1989.

[Sg2]   C.L. Siegel: *Topics in Complex Function Theory, volume I*,
             Wiley-Interscience 1971.

[Sh1]   N. A. Shah: Limit distributions of polynomial trajectories
             on homogeneous spaces, *Duke Math. J* **75**, (1994), 711–
             732.

[Sh2]   N. A. Shah: Invariant measures and orbit closures on ho-
             mogeneous spaces, in: *Proc. International Colloquium on
             Lie Groups and Ergodic Theory*, Tata Institute of Fun-

damental Research, Narosa Publishing House, New Delhi 1998, 229–271.

[Si] Y.G. Sinai: *Topics in Ergodic Theory*, Princeton, 1994.

[St1] A.N. Starkov: Ergodic decomposition of flows on homogeneous spaces of finite volume, *Math. USSR Sbornik* **68**, (1991), 483–502.

[St2] A.N. Starkov: On multiple mixing of homogeneous spaces, *Russian Acad. Sci., Dokl. Math.* **48**, (1994), 573–578.

[St3] A.N. Starkov: New progress in the theory of homogeneous spaces, *Russian Math. Surveys* **52**, (1997), 721–818.

[Va1] V. S. Varadarajan: Groups and automorphisms of Borel spaces, *Trans. Amer. Math. Soc.* **109**, (1963), 191–220.

[Va2] V. S. Varadarajan: *Lie Groups, Lie Algebras, and their Representations*, Springer Verlag, 1984.

[Ve1] W.A. Veech: Weakly almost periodic functions on semisimple groups, *Math. Z.* **88**, (1979), 55–68.

[Ve2] W. A. Veech: Unique ergodicity of horospherical flows, *Amer. J. Math.* **99**, (1977), 827–859.

[Vi] N. Ya. Vilenkin: *Special Functions and the Theory of Group Representations*, Transl. of Math. Monographs, American Mathematical Society, 1968.

[Wa1] G.L. Watson: On indefinite quadratic forms in five variables, *Proc. London Math. Soc. (30)* **3**, (1953), 170–181.

[Wa2] G.L. Watson: On indefinite quadratic forms in three or four variables, *J. London Math. Soc.* **28**, (1953), 239–242.

[War] F. Warner: *Foundations of Differential Manifolds and Lie groups*, Springer Verlag, (1983).

[We] A. Weil: *L'Intégration dans les Groupes Topologiques et ses Applications*, Hermann, 1965.

[Wi] D. Witte: Rigidity of some translations on homogeneous spaces, *Invent. Math.* **81**, (1985), 1–27.

[Wl] P. Walters: *An Introduction to Ergodic Theory*, Springer Verlag,1982.

[Wr1] G. Warner: *Harmonic Analysis on Semi–simple Lie Groups I*, Springer Verlag, 1972.

[Wr2] G. Warner: *Harmonic Analysis on Semi–simple Lie Groups II*, Springer Verlag, 1972.

[Zi] R.J. Zimmer: *Ergodic Theory and Semisimple Lie Groups*, Birkhäuser, 1984.

Index

algebraic measure   185
Ad -unipotent   184
amenable group   32

Bernoulli-shift   8
Birkhoff's ergodic theorem   17
Borel fundamental domain   48
Bruhat decomposition   110

Cartan decomposition   67, 82
Cartan involution   67
complete Riemannian manifold   68
complete subgroup   12, 154
counting problem   93
cusp   56

determinant   151
Dirichlet region   44
diverging to infinity   23
dual group   6

elliptic element   53
equidistribution   13, 95, 129, 186
ergodic action   2
ergodic decomposition   31
ergodic hypothesis of Bolzmann   5
essentially invariant   2
extreme point   30

finite measure   1
flow   11
free action   66
Fuchsian group   43
fundamental domain   44

Gauß–Bonnet formula   50
Gauß's circle problem   93
Generic point   32
geodesic   38
geodesic flow   36, 57
geometrically finite   51

Haar measure   2
Hedlund's minimality theorem   116
horocycle   112

horocycle flow   112
Howe–Moore's theorem   81
hyperbolic element   53
hyperbolic plane   38

indefinite form   162
invariant measure   2
irrational form   161
Iwasawa decomposition   110, 140

Killing form   67
Kronecker's theorem   12

lattice   42
Liouville's theorem   9
locally finite measure   2
locallly symmetric space   65

Mahler's criterion   147
Margulis' lemma   150
matrix coefficient   23, 81
Mautner lemma   61, 83
minimal action   114
minimal flow   12
mixing of all orders   98
modular group   45
Möbius transformation   39
Moore's ergodicity theorem   89, 90
Moore's duality theorem   91

no compact factors   81

Oppenheim's conjecture   162

parabolic element   53
Poincaré's disc   42
Poincaré's recurrence   8
Poincaré's upper halfplane   38
positive definite form   148
primitive vector   162
proper group action   65
properly discontinuous   43

quadratic form   148
quasi-invariant measure   2

Raghunathan's conjecture·   184
rank of a symmetric space   74
rational form   161

regular measure   2
Riemmanian structure   37
Riemannian symmetric pair   63
Riemannian symmetric space   62
Riesz' representation theorem   31

Siegel sets   140
Siegel's theorem   51
σ-finite measure   1
simple factor   81
strong law of large numbers   19
strong mixing   21, 23
symmetry   62

tending to infinity   23
totally non-compact   90

uniform lattice   43
uniformly continuous   123
uniform distribution   186

unimodular group   42
unimodular lattice   146
unipotent one-parameter group   150
unique ergodicity   32
unit tangent bundle   36
unitary representation   13

vanishing at infinity   23
vertex   49
vertex at infinity   49
von Neumann's ergodic theorem   15

weak mixing   25
Weil formula   42
Weyl's equidistribution theorem   13

Printed in the United States
By Bookmasters